SCIENCE AND THE ANTE-BELLUM AMERICAN COLLEGE

Memoirs of the
AMERICAN PHILOSOPHICAL SOCIETY
Held at Philadelphia
For Promoting Useful Knowledge
Volume 109

SCIENCE
AND THE
ANTE-BELLUM AMERICAN COLLEGE

STANLEY M. GURALNICK

Colorado School of Mines

THE AMERICAN PHILOSOPHICAL SOCIETY

Independence Square • Philadelphia

1975

Library of Congress Catalog Card Number 75-12219
International Standard Book Number 0-87169-109-4
US ISSN 0065-9738

CONTENTS

Preface

THE period from 1820 to 1860 was one of great social and cultural ferment in America. Called the age of the common man for its sporadic contribution to growth in popular political participation, it constantly witnessed the increasing division of labor characteristic of an incipient industrial economy, the proliferation of new institutions, and the introduction of changes to existing ones. An important, though infrequently discussed example of such institutional change is the awakening which took place within the colleges of America, resulting, among other things, in the elevation of science to an unprecedentedly important, almost dominant position in the structure of the liberal arts program.

Although to some extent science had always, even so far back as the seventeenth century, enjoyed a place in the American college curriculum, earliest instruction had been minimal. Indeed, it was likely to be confined to the memorization of the principal scholastic commentaries upon the physical world and to the performance of simple arithmetical operations in mathematics—activities which could only become increasingly removed from the concerns of actual practitioners in the developing scientifically revolutionary age. In addition, the total effort then expended in the area of scientific instruction amounted only to an insignificant fraction of the current classical curriculum, at that time overwhelmingly dominated by the ancient languages.

By the mid-eighteenth century, the growing collegiate interest in scientific scholarship which the Enlightenment had aroused found expression in the more frequent if still only occasional appointment of science professors to college faculties. But the overall effect of their presence, even in the few instances when such professors were also able scientists, was less enduring than their fame might suggest; for each was burdened with instruction in mathematics and possibly other subjects as well. Moreover, it was not atypical for the college of the late eighteenth and even the early nineteenth centuries to experience long periods when no science professor graced its faculty (instruction being offered by tutors, if at all), when no material aids for instruction were available, and when no expression of concern was voiced for any of these conditions.

Somewhat ironically, the American college would not become fully aware of how backward its curriculum had been rendered by the scientific discoveries of Europe until the nationalistic era of intellectual isolation which succeeded the War of 1812. As this study will demonstrate in some detail, it was only then that Americans, while not desiring to emulate the national scientific or educational institutions of Europe, first began to show sustained interest in upgrading the level of scientific instruction offered within the educational institutions of their own country. And it was then, too, that the college, in response to such sentiment, began to devise plans for the addition of science courses to the standard requirements for the bachelor of arts degree, thereby incidentally diminishing the emphasis formerly placed on the older subject matter of that program, requiring a greater division of labor within the faculty, and necessitating the purchase of the equipment essential to the teaching of science at the nineteenth-century level.

The anxiety the college felt about its existing scientific inadequacy was heightened by the fact that many collegiate institutions had already failed for lack of popular support, while others (some among the twentieth century's most prosperous) threatened to join them. A transformation of the college program to include more science was expected, by innovative educators, to make the sparsely attended colleges more appealing to the country's youth. In addition, leading theorists expressed hope that the new additions would mute growing criticism of the college's program as totally irrelevant to the bustling world of a materially expanding America.

Accordingly, during the 1820's, in an effort to obtain both intellectual sophistication and wider public appeal, the college initiated debates over the form that curricular change *vis-à-vis* science should take. Although a few educators used these occasions to urge the creation of true universities offering intense graduate training like those in Germany, such suggestions found little support in an America not yet ready to replace its democratic belief in the appropriateness of a single program of study for all students with a foreign scheme of higher study for the intellectually elite. Similarly, the opinion of those who urged the establishment of vocational programs, designed to make the college attractive to select groups seeking to learn a trade, was discouraged. Rather, the most common result of these evaluative surveys was the urging of the college to act upon its yet amorphous belief that the number of subjects required of all students would have to be increased by the addition of more mathematics and science to the standardized bachelor of arts program.

And so, whether their financial support came from a modest increase in the size of their student bodies, from the public funds which variously found their way into private institutions, or from the enlarged benefactions of the private philanthropy that characterized an age of growing personal fortune, those colleges which survived the late 1820's set themselves on the distinguishing course of the ante-bellum decades—diffusion of scientific knowledge through scientific expansion. Their new appeal to students, parents, and public would be based on a more modern curriculum taught by competent and permanent professors, rather than by transient tutors. Thus, collegiate employment was offered to a growing cadre of young science teachers; while joint professorships, such as those in mathematics and natural philosophy, were now divided into two and more positions as more subjects entered the program of required study.

Overall, the changes that marked this expansion within the college are as easy to outline as they were profound in their consequences. Where the whole of mathematics and science instruction within the American college curriculum of 1800, for instance, had been contained in two books (and those of doubtful quality), that of 1825 was taught from no less than four, and that of 1850 from a minimum of ten. Where scientific subjects had commonly engaged one professor per school in 1800, by 1830 there were usually two, and by 1860 four, with occasional instances of scientists making up over half of an individual college's faculty. A tenfold increase in the total number of American science professors recruited from 1828 to 1860 reverberated even beyond the college walls, as these new professors became active members of the national scientific community insofar as it existed. And total material resources for scientific instruction, which had been valued at a few hundred dollars in 1820, grew by 1860 to many thousands, not including the cost of the separate laboratory buildings in which they were housed—and this even at colleges which were never to become universities. Identifying 1825 as the beginning of "decided or marked improvement in colleges," one observer at mid-century remarked of the "great changes in the course of study" that had taken place over the prior fifty years:

> There have been added to the list of college studies chemistry, geology, mineralogy, botany, political economy, and the modern languages; while the departments of mathematics, intellectual and natural philosophy, have been greatly enlarged. . . . In 1801, the quantity of apparatus was very small, and the experimental lectures of the professors very meagre; now, almost every college is able to illustrate all the principles of the sciences very fully and to show their practical application.[1]

more resources

Scientific activity, in fact, proved to be the preeminent concern of this difficult period and the largest drain on the resources of those colleges that survived. A professor at the University of Mississippi, for instance, estimated in 1858 that, while his school's library resources were worth $10,000, the natural history collections were worth $4,000 and the apparatus in natural philosophy $15,000.[2] Moreover, the spirit of urgency which accompanied these changes cannot be overestimated. By the 1830's, it seemed that wherever there was a college, the "philosophical apparatus [was] the thing which demanded the greatest attention."[3] Thus the Board of Trustees of Transylvania, the first trans-Appalachian college, learned that five thousand dollars spent on scientific equipment would "be more promotive of the welfare of the institution than any single step that could be taken," as it would

---

[1] Emerson Davis, *The Half Century* (Boston, 1851), p. 7. A valuable modern aid for the analysis of social science developments in the changing curriculum is Wilson Smith, *Professors and Public Ethics* (Ithaca, 1956).

[2] MS, F.A.P. Barnard to D.L. Swain [President of University of North Carolina], Nov. 15, 1858, University Papers, University of North Carolina Archives.

[3] MS, George Wyllys Benedict to J.J. Robertson, March 5, 1836, University of Vermont Archives.

help both in keeping the present science professor and in enticing another to join him.[4]

Yet by the 1840's, the then two-decades-old procedure of enlisting professorial aid in adding science after science and subject after subject to the standard curriculum seemed to result in a new crisis. The student opinion that had so often demanded new courses became equally insistent upon not studying them. With the gap considerably narrowed between the intellectual levels of the college textbook and the scientific journals, some professors began to offer the rationale of science-for-its-own-sake as a substitute for the doctrine of mental discipline that had always before defined the college's approach to knowledge. Others tried the college's capabilities by asking it to show cause how the intellectual utility it provided was useful in satisfying society's more tangible aims. And realistically minded reformers, meanwhile, began to question the value of continued scientific course accretions, asking where the college could go now that more than twenty subjects made up the total curriculum.

All answers to this question recognized that superficiality would be the inevitable result of trying to satisfy every new need by cramming still more courses into a curriculum of fixed length, and thus alternatives to that single method of accommodation were proposed. A number of public spirited manufacturers thought the college should arrange to train better mechanics, some spokesmen for scientific agriculture urged that it train better farmers, and still others entertained visions of graduate programs beyond the circle of the liberal arts or, more modestly, of college course electives. The purpose of this study, however, is not to trace later, localized nineteenth-century developments from their origins in these mid-century curriculum demands, but to provide a detailed examination of the changes that the college of the traditionally decentralized system of American higher education had already undergone before them.

In the second quarter of the nineteenth century, where our attention is focused, American democracy required a greater diffusion of scientific enthusiasm among many segments of

[4] MS, B.O. Peers to Board of Trustees, [?] 1833, College Papers, Transylvania College Library.

the population before the environment could offer enough nourishment for special forms of culture, like those mentioned above, to grow; and the classical college helped fill this need. Thus, potential contributors to scientific knowledge themselves benefited from the amiable acquaintance with scientific discipline that the college provided and from the variety of social models, including that of the overworked college professor, which it actively entailed. Without the resources to support major research laboratories and relaxed teaching loads as they were known to a later age, the college nevertheless made a significant social response during the period discussed. Methodologically, the emphasis developed here assumes that merely to recount successful direct anticipations of later university experiences—the more common basis of evaluation in educational historiography—is not as historically important as to understand these indirect preparations, born as much out of contemporary need as future desire.

Since the usual historical presentation of the college as a somewhat stagnant structure has made no use of the large body of evidence that suggests an alternative view, I have based my demonstration of the dynamic relationship that actually existed between science and college on a variety of previously unexplored primary sources. The study concentrates principally on fifteen northeastern colleges which together provided the country with the majority of its college graduates in 1830,[5] although some additional supportive evidence gathered from several southern institutions is included. Individually these colleges represent diverse environments and denominational influences; and while some went on, from their common experience, to become great universities, others maintained themselves as comparatively rural collegiate institutions. Thus, the sample is presumably adequate for the presentation of a well-balanced and representative picture.

Although it is not my intention to provide a detailed analysis of each scientific discipline, some subject matter must be inspected in order to support the thesis. I cannot, for example, expect to convince the reader that the mathematics

[5] Amherst, Bowdoin, Brown, Columbia, Dartmouth, Dickinson, Harvard, Middlebury, University of Pennsylvania, Princeton, Union, University of Vermont, Wesleyan, Williams, and Yale.

curriculum of 1860 was far more advanced than that of 1820 without explicity showing what kind of problems the students of each period were taught to solve. The result of such exposition is not to be regarded as a history of mathematics in America during the period, however, but rather as a type of evidence indicating that the changes I am asserting really took place—a source of confirmation more reliable than the rhetoric of commencement addresses, or deceptive catalog descriptions.

Similarly, for other topics of interest, no attempt has been made at encyclopedic completeness. There seems to be no necessity, for instance, to catalog the name and activity of each student society or every penny spent on scientific equipment in all the colleges, for these serve only as examples of the many forms that interest in science assumed.

The principal sources for this book have been the archives of the colleges studied—primarily manuscript material, such as trustees' records, student notebooks and diaries, faculty reports, examinations and correspondence. Only very limited use has been made of standard college or educational histories. The chief printed sources are nineteenth-century periodicals, textbooks, catalogs, orations, and miscellaneous college publications. Together, these diverse materials reveal that, in only a few decades, the college learned to struggle for a rapport between curriculum and social needs that the older, scholastically derived programs could not provide in an increasingly complex and industrial age; that science came to permeate the lives of all these institutions; and that their own intellectual accommodation to the demands of science performed the function of cultural diffusion upon which scientific schools and other developments were later dependent.

The research for this book could not have been done without the assistance of librarians and archivists at the following institutions of higher learning: Amherst College, Bowdoin College, Brown University, Columbia University, Dartmouth College, Dickinson College, Duke University, Harvard University, Middlebury College, University of North Carolina, University of Pennsylvania, Princeton University, University of Rochester, Transylvania College, Union College, University of Vermont, University of Virginia, Wesleyan University, Williams College, and Yale University.

I also want to acknowledge the assistance received at the following non-university libraries in Philadelphia and Washington: the Library Company of Philadelphia, the Franklin Institute, the Historical Society of Pennsylvania, the American Philosophical Society, the Library of Congress, and the Smithsonian Institution.

The book began at the University of Pennsylvania as a doctoral dissertation entitled "Science and the American College, 1828-1860" (1969). There I received assistance and encouragement from Professors Russell McCormmach (History of Science), Arnold Thackray (History of Science), and Murray Murphey (American Civilization). But I am especially indebted to my dissertation supervisor, Professor Charles E. Rosenberg (History), for his expert guidance and for his patience in helping a naive graduate student grow to the task.

I gained many insights for revising the original dissertation from Professor Rosenberg and from Dr. Nathan Reingold, editor of the Joseph Henry Papers, Smithsonian Institution. The final version benefited from the incisive criticism of Professor Harold I. Sharlin of Iowa State University and from the editorial assistance of the American Philosophical Society.

I am indebted to the United States government for an NDEA Title IV fellowship, 1966-1969, and to the Graduate School of the University of Pennsylvania for supplementary research funds to complete the dissertation. During the year 1969–1970 I was privileged to be Visiting Postdoctoral Research Fellow at the Smithsonian Institution where I enjoyed the Institution's financial and intellectual support to revise and enlarge the dissertation.

Family and friends have endured my work-connected frustrations for years. Finally, I thank my wife, Elissa, for her criticism, encouragement, and advocacy.

Boulder, Colorado S.M.G.
March, 1975

I. Background to Science and the College in the Nineteenth Century

Western intellectual tradition was markedly altered in the eighteenth century on the heels of the scientific revolution. Increases in the amount of scientific knowledge and communication, not to mention corresponding adaptations in the philosophical tradition of the Enlightenment, inevitably led to structural changes in the educational institutions charged with the formal training of the literate minorities.

Thus, in the latter half of the century, German universities followed Halle and Göttingen in providing expanded opportunities for the study of biology and physical science in addition to the professional subjects of medicine, law, and theology bequeathed by the Middle Ages and the classical curriculum of Renaissance humanism. English universities broke with their long-standing classical tradition in the adoption of the English language as a medium of discourse. And the Scots were so impressed with scientific developments that by 1811, nineteen out of thirty-three professors at the University of Edinburgh taught either science, mathematics, or medicine. Napoleon, even more strikingly, sought to infuse national culture with science through the creation of a national system of education culminating in special scientific-technical schools.

However many years remained before the universities of Europe would assume their science-dominated guise of more recent times, the eighteenth century had witnessed a rapid assimilation of modern science into university thought, perhaps in response to Europe's having glimpsed the problem of relating education to a society destined to be continuously transformed by scientific culture. It is a measure of the universities' importance that even those scientists who berated them for not assisting in their scientific pursuits looked to them to create scientific careers and understanding for others.

How well each country's university system succeeded in meeting the challenge of scientific culture is a subject that still provokes violent disagreements, for there is an inherent difficulty in evaluating university contributions to scientific advance apart from other forms of intellectual support, as well as in correlating scientific productivity with a given university reform. But while we may argue over the problems of which nation taught more science in 1790 or better fostered its advances in 1810, we should not lose sight of the important fact that Western education was everywhere changing, and that it was the European and American colleges, not those of China, Africa, or Latin America, that were struggling with the age's challenge to their dominant intellectual traditions.

America, despite all its colonial dependence on Europe in matters of taste, tradition, and affectation, exhibited throughout the eighteenth century a native scientific awareness and a growing sense of the importance of science to the future of its continent. Although the colleges of colonial America have often been mistaken for seminaries designed for the training of ministers in the complexities of theology and in the duties of their post, closer study of these schools has created a new scholarly recognition of their increasingly secular orientation and, in particular, of their own minor revolution in adhering to Enlightenment standards in the cultivation of the reason. Admittedly, these small colleges of a few dozen pupils, two or three traditional scholars, and perhaps a single science professor in years of good tuition can hardly be compared to the University of Edinburgh in resources or accomplishment. But such disparity between the best institutions of Europe and the struggling colleges of new America is hardly an excuse for neglecting significant transformations in the latter.

With its own origins in the universities of medieval Europe and its inspiration for well over a century distinctly provided by European intellectual precursors, the American colleges had already undergone a long history of development by the early nineteenth century. In the seventeenth and eighteenth centuries, when they were being established, they appeared to differ greatly from one another in particulars. Scientific instruction, for instance, might in any given year be nearly sophisticated at one while almost non-existent at another. However, despite these superficial differences that we shall see existed, there was a basic similarity developing among all these institutions, the oldest and the newest; and by the late eighteenth century, they had all become distinctly American institutions. This is the story of their earliest formative years.

The history of American higher education begins in the fall of 1636 with the legislative acts establishing Harvard College. The purposes and methods of seventeenth-century education differ as markedly from our own as did the century itself. After two hundred years of Renaissance scholarship, the age was still intellectually closer to the medieval world than to the post-Newtonian world. The great theme of all learning was still the Christian drama. The little world with the earth at the center was supposed to have begun in 4004 B.C. Only a few years before, Galileo had been humiliated by the Roman Inquisition for his support of the Copernican thesis.

Still widely discounted almost a century after its formal announcement in 1543, the Copernican system was only one of a number of major sixteenth-century developments transforming the Western world. Exploration, mercantilism, technological improvement, and the unanticipated social upheavals of the Protestant Reformation all contributed to the dissolution of the medieval world view. And they were equally responsible for bringing the English Puritans to New England.

The Puritans provide a very significant chapter of the early intellectual development of this country. From the monumental labors of Perry Miller, we have come to appreciate the complex texture of Puritan life in the wilderness, to be conscious of social change in their increasingly secular communities, and to be more sympathetic to their archaic defenses of church polity and doctrine. From Miller's work has come the realization that the Puritans did not establish Harvard as a church school for training clergymen in systematic theology. Their purpose was to train learned Christian citizens. For the Puritans, "a theological seminary had no more meaning . . . than an engineering school. They did not distinguish between secular and theological learning; and they believed that the collegiate education proper for a minister should be the same as for an educated layman."[1] It was only the simplicity of their economic and social life that demanded that over half of these good, educated Christian men become Congregational ministers. The ideal of educating Christians as classicists rather than as free thinkers or productive scholars left its mark upon the American college for at least a century.

Harvard adopted the subject matter of the Middle Ages without the earlier flexibility. Even in the fourteenth century, the recitations of Aristotelian physical theses at Oxford and Paris were accompanied by numerous scholarly tracts, and older scholars created a spirited academic community. In seventeenth-century America, however, the statement of physical theses at commencement was a sterile display for teenagers, often nothing more than mere definitions.[2]

The teaching staff at Harvard throughout the century consisted of the president and young inexperienced tutors waiting for a pulpit. Tutors usually supervised particular classes through the entire three- (later four-) year curriculum. There were no academic departments and often not very competent tutors.

It is a fair assumption that no one could have learned very much science in the seventeenth century. Two days of the week were spent studying Greek and Semitic languages. One day was devoted entirely to rhetoric and another to divinity. This left two days to read philosophy, logic, ethics, politics, and physical science in Latin textbooks. The only career open to young graduates of scholarly tastes was the ministry.[3] There was no formal training for the professions of law or medicine.

The mathematics instruction demonstrates a lack of interest in contemporary physical science. In the three-year curriculum of 1642, the only provision for mathematical instruction

---

[1] Richard Hofstadter and Wilson Smith, *American Higher Education: A Documentary History* (Chicago, 1961) I: p.2. It may be helpful to emphasize that a prescribed course of theological instruction was not even a conception of the early medieval university but took form in the latter sixteenth century as one educational response to the Reformation. See Charles Homer Haskins, *The Rise of Universities* (Ithaca, 1957), p.34. The arts curriculum of the English universities, which served as the first model of the American liberal arts program, itself contained little religion. See Craig Thompson, *Universities in Tudor England* (Washington, 1959), p.2.

[2] Samuel Eliot Morison, *Harvard College in the Seventeenth Century* (Cambridge, 1936) 1: p.159ff. James J. Walsh, *Education of the Founding Fathers of the Republic* (New York, 1935).

[3] Morison, *ibid.* 2: p.557.

is in the third or senior year, when two hours of recitation per week are devoted to arithmetic, geometry, and astronomy.[4]

At the end of the seventeenth century, there remained no change in the Harvard mathematical curriculum. Mathematical studies were still considered as "subjects fit for mechanics rather than men of learning."[5] The level of instruction was reflected in mathematical theses on such erudite topics as "Nine notes or figures are the mathematical alphabet" and "A cipher itself signifies naught."[6] Even in 1725, senior students had to review their childhood arithmetic for a month before they could study only the first five books of Euclid. The first three years still provided no mathematical instruction. Algebra and trigonometry entered the curriculum only in 1726. Florian Cajori found that "ninety years after the founding at Harvard, the mathematical course was essentially the same as at the beginning."[7]

The first inroad against memorization and recitation of the Aristotelian physical categories was made by Charles Morton, who arrived in 1686, "perhaps the most accomplished scholar that came to New England in the colonial period."[8] Educated at both Cambridge and Oxford, Morton was keenly aware of the ferment produced by the new philosophy and experimental science. He was at Oxford when the "Invisible College," predecessor of the Royal Society, moved to Oxford from London in 1648. And he was personally acquainted with many future Fellows, including Thomas Sprat, John Wallis, and Robert Boyle.

After the Restoration and the exclusion of Puritans from Oxford and Cambridge, this capable teacher opened one of the most successful of the dissenting academies, which, by devoting time to modern languages and science, provided a more modern education than did the ancient seats of English learning.[9] Sometime between 1680 and 1685 Morton prepared for students his manuscript manual, soon to be called the

---

[4] Morison, *ibid.* 1: p.141.

[5] Morison, *ibid.* 1: p.208.

[6] Morison, *ibid.* 1: p.209.

[7] Florian Cajori, *The Teaching and History of Mathematics in the United States* (Washington, 1890), p.22.

[8] Edward Eggleston, *The Transit of Civilization from England to America in the Seventeenth Century* (New York, 1901), p.45.

[9] On Morton's innovative procedures in the teaching of science in England and on the dissenting academies in general, see Joe William Ashley Smith, *The Birth of*

Compendium Physicae. Of some hundred pages in length, this manuscript in English would prove more useful to young students than the lengthy compendia of the late seventeenth century.

Finding Royalist persecution of the academies intolerable, Morton responded to a letter from Increase Mather requesting him to come to Boston. The latter suggested that Morton might be made president of the College, but political circumstances surrounding the revocation of the Bay Charter made that impossible. Instead, Morton was installed as head of John Harvard's old church, the First Church of Charlestown. He became a leading New England intellectual and actively engaged in the administration of Harvard affairs and in such civic problems as the famous witch hunts. For a time, he gave private instruction in science but was not a member of the faculty.

His textbook, however, helped Harvard pull "through the bogs of scholastic physics."[10] Hornberger has called the *Compendium* "a microcosm of men's speculation about the external world from the time of Empedocles down to the eve of Newton's *Principia.*"[11]

The *Compendium* presents several bodies of thought. The first, already familiar to students, is an elaboration upon the assorted works of Aristotle. The second is a presentation of Bacon and especially of Descartes as systematic opponents of the Stagirite. The third, and most important, is the discussions, often illustrated, of seventeenth-century science. The researches of Galileo, Boyle, and others are presented directly from their own and from other contemporary works.[12] There is a lengthy discussion of cosmologies including the Tychonic and Copernican.[13] There are discussions of the properties of

Modern Education: The Contribution of the Dissenting Academies, 1660-1800 (London, 1954), pp.56-62, 246-250, and *passim.*

[10] Morison, *op.cit.* 1: p.251.

[11] In introduction to *Charles Morton's Compendium Physicae* (Boston, 1940), p.xxxvi. The range of Morton's interest is suggested in N. J. Seeley, "Two Autograph Manuscripts of Charles Morton in the University of London Library," *Journ. Royal Inst. Cornwall,* 5 (1965): pp.100-108.

[12] Morton had been a pioneer in England in incorporating the early volumes of the *Transactions of the Royal Society* into his course preparations, and the *Compendium* reflects his continuing interest in contemporary science.

[13] *Charles Morton's Compendium Physicae,* Publication of the Colonial Society of Massachusetts, Collections 33 (Boston, 1940): pp.24-25.

air, directly from Boyle. Pneumatic devices and thermometers, all yet unfamiliar to Harvard students, are illustrated. The problem of the tides, including Galileo's contradictory analysis, is discussed.[14]

Although much of the *Compendium* was obsolete when it was first read and completely outdated when Newton published his monumental *Principia* a few months later, there was not to be another change in the Harvard science curriculum until 1728. But in a Puritan world which previously had only scholasticism, Protestantism, and Renaissance Humanism to draw upon, the *Compendium* presented a new way of thinking. The impact of Morton and his manual was immediate. The *theses physicae* of 1687 contain not only *"Generation est principium Corruptionis,"* but also *"Gravitas est vix attractrix terrae."*[15]

Thus, at the conclusion of the seventeenth century, mathematics and science played a small role in the Harvard curriculum. Neither had ever been completely excluded but neither had received any real encouragement. Harvard could not be favorably compared with the invigorated universities of Europe, where early post-Newtonian lecturers would soon incorporate some form of the master's researches into the intellectual milieu.

Yale, the second college in America, founded in 1701 and settled in New Haven in 1718, was organized by Harvard men, all of whom had graduated before Morton's *Compendium* entered the curriculum. Abraham Pierson, first rector and author of a scholastic manual for Yale students (since lost), had left Harvard in 1667 when Magirus's *Physiologicae Peripatectica* (1597) was still used. Thus, Yale's science curriculum was even more anachronistic than Harvard's. It consisted of Rector Pierson's notes, other scholastic texts, and preparations of "Technologia." Pierson himself had "undoubted faith in the theory that the sun rolled around the earth,"[16] and "Technologia" did not even deal with such topics of recent concern.

"Technologia" was a collection of *theses technologiae,* derived from the early seventeenth-century writings of Alexander

[14] *Ibid.,* pp.63-65.

[15] Morison, in preface to above, p.xxiii.

[16] Edwin Oviatt, *The Beginning of Yale, 1701-1726* (New Haven, 1926), p.422.

Richardson and William Ames. Richardson had established the "rule of encyclopedia," a philosophy connecting the liberal arts in a "circle of seven sections of which the center is God."[17] Using this scheme, Ames elaborated the nature and ends of all the arts and intellectual faculties. This collection of "simple, straightforward, declarative axioms" served as the framework of all Puritan philosophy, psychology, and theology as it had developed from the philosophy of Petrus Ramus and as it was also known at Harvard.[18]

"Technologia" provided Yale with its only available intellectual perspective of science, which was not considered as a collection of disciplines or techniques but as a highly intellectualized arc of the arts circle. Science could never exist either for its own sake or for the purpose of filling the needs of society: it had to be constantly related to the whole of knowledge and ultimately to God. The objects of the physical world could never be entities for individual study. They were not only objects of God's divine physical and logical order, but also inextricably involved in His moral plans.

Samuel Johnson, later the first president of Columbia, prepared as a senior at Yale in 1714 a "Synopsis Philosophiae Naturalis"[19] and an "Encyclopedia of Philosophy."[20] The former, a series of questions and answers such as appeared in books "100 or 150 years old" that "the first settlers of the country [had] brought with them 70 or 80 years before," according to Johnson, considered such scholastic problems as the origin of qualities, the nature of angels, and the theological implications of the eternity of the earth.[21] The collection of

---

[17] Perry Miller, *The New England Mind: The Seventeenth Century* (Boston, 1961), p.160.

[18] *Ibid.*, p.159. Ramus was the Protestant philosophical hero of the sixteenth century who, in a 1536 thesis, declared "all that Aristotle said is false." Nicknamed "usuarius" because of his desire to find a modern use for everything, Ramus thought that Virgil might be helpful in teaching modern science and Cicero in astronomy. He made French and Latin translations of the first six books of Euclid and other mathematics books, but his fame in Puritan circles rests upon his anti-Aristotelian logic. He was decapitated in the St. Bartholomew Day Massacre (1572). See Frank Peirrpont Graves, *Peter Ramus and the Educational Reformation of the Sixteenth Century* (New York, 1912), pp.26, 57-59, *passim*.

[19] Printed in Herbert and Carol Schneider, eds.; *Samuel Johnson, President of King's College: His Career and Writings* (New York, 1929) **2:** pp.23-54.

[20] "Technologia sive Technometria; Ars Encyclopaidia manualis ceu philosophia," also translated and printed in *ibid.*, pp.55-186.

[21] Quotations from Johnson's autobiography, *ibid.* **1:** p.6.

1271 sterile definitions which constituted the "Encyclopedia" was the crowning achievement of his college career and showed the full extent of his understanding of mathematics and the physical world. Neither reveal any knowledge of Copernicus nor of the physics of the sixteenth or seventeenth century. We can imagine why he expressed such amazement and delight the following year when Newton and Locke were introduced in Connecticut.

So far as mathematics was concerned, the curriculum at Yale was even more primitive than that at Harvard in its early years. Students studied "common arithmetic and a little surveying" in the last year of college without any formal geometry.[22]

Meanwhile, in Virginia, the colonists, who had discussed the founding of a college as early as 1617, secured a charter for William and Mary in 1693. Instruction began in 1710 at this institution, whose importance has been obscured by the fact that so many of its early records have been destroyed by fire. The College was founded by Oxford rather than Cambridge men and as such came under the immediate influence of the revised Oxford curriculum, which provided for a classical or grammar school, a post-graduate divinity school, and a college or philosophical school. One of the two college divisions was devoted entirely to mathematics and natural philosophy.

There is no record of how the early program operated, but an actual professorship of mathematics and natural philosophy was established in 1712. Reverend Hugh Jones, educated in England, occupied the position from 1717 to 1729 and was the first professor of science in America. The most competent of the nine men who taught science at William and Mary before the Revolution was William Small, who had the position from 1758 to 1764. Thomas Jefferson studied under Small and acknowledged him as the source of his lifelong interest in science. A large number of Virginia patriots received their college training at William and Mary where politics and modern languages were also found in the curriculum.

From 1710 to 1740 the three established colleges received the impact of the scientific revolution. A significant increase in

---

22 Cajori, *op.cit.*, p.28.

the number of non-academic scientists, particularly natural historians, and the diffusion of the many non-technical and philosophical derivatives of Newtonianism contributed to an excitement, further enlarged by the election of seventeen colonials as Fellows of the Royal Society.

There probably was no copy of Newton's *Principia* in the colonies until 1708, and even then no one in America was capable of understanding it.[23] The more easily digested *Optiks,* by which the memory of Newton was consecrated to so many of the eighteenth century, was not yet popular. Although the Harvard library catalog of 1723 does not even list a copy of the *Principia,* post-Newtonian literature, in textbooks and in the *Transactions of the Royal Society,* found its way into the colleges. In 1715 the famous collection of books solicited in London by Connecticut agent Jeremiah Dummer reached Yale. Included were the *Optiks,* the second edition of the *Principia,* donated by Newton himself, and Locke's *Essay on Human Understanding.* Newton and Locke could spawn varying conclusions about fundamental epistemological questions, but they had the singular effect of stimulating great numbers to ever widening mental exertions. They were generating too much excitement to permit the former complacency to permeate the college much longer. Immediately, Samuel Johnson set out to learn the new philosophy and Newton's fluxions as well.[24] As tutor at Yale, Johnson helped interest students in Newton and Locke.

But even for the second generation of Yale students "much of the antiquated, supernatural rubbish of an earlier scholastic time was still taught, and indeed remained for many a year the conservative teaching of the College."[25] In 1719, when Jonathan Edwards was a student at Yale, the 1653 *Institutio Astronomica* of Pierre Gassendi was still the recommended astronomy text.[26]

During the presidency of Thomas Clap (1739-1756), a gifted

[23] On the lack of scientific expertise in America, see Thomas Brattle's 1705 letter printed in Frederick Kilgour, "The Rise of Scientific Activity in Colonial New England," *Yale Jour. Biology and Medicine* 22 (1949): p.126.

[24] See Johnson's autobiography in H. and C. Schneider, *op.cit.* 1: pp.6-9 and the Schneiders' own remarks, 2: pp.6-7.

[25] Oviatt, *op.cit.,* p.422.

[26] Louis F. Snow, *The College Curriculum in the United States* (New York, 1907), p.47.

scientific amateur, the entire Yale class received its first taste of fluxions. In 1734 Yale received a set of surveying instruments, telescope, microscope, and barometer from an English merchant. To what use these were put is uncertain, but in Clap's program of 1742-1743 we find the study of Euclid's *Elements*, Gravesand's recent natural philosophy text, and a modern astronomy.[27] By 1746, Clap had established the elements of arithmetic as an entrance requirement at Yale, the first crack in the Latin and Greek hegemony in over a hundred years of American college experience.

Harvard, too, was undergoing similar change. In 1726 Cotton Mather, the old conservative who gained international recognition for his espousal of smallpox innoculation, advised his fellow ministers: "What we call NATURAL PHILOSOPHY, is what I must encourage you to spend much more time in the study of."[28] In 1727 the famous Hollis Chair of Mathematics and Natural Philosophy was established at Harvard and filled the following year by Isaac Greenwood. Greenwood, who had graduated from Harvard in 1721 and had gone to London to study science, performed admirably on his return. But the special distinction of the Hollis Chair was created by John Winthrop, who taught from 1738 to 1779. Winthrop was a first-class mathematician and astronomer who also had a special gift for instruction. He used calculus in his astronomy lectures beginning in 1751 and then led the first college-sponsored scientific expedition in 1761 when he went to Newfoundland to observe the transit of Venus. Measurements of this kind were part of the international effort to verify and extend Newton's laws.

With the cooperation of President Holyoke (1737-1769), the last traces of Aristotle disappeared from the course content in physics, metaphysics, and logic. "Over a period of twenty-five years so many obsolete books were replaced by new works that the undergraduate course at the end of Holyoke's regime had little in common with that of Leverett's day."[29] The change in mood that had swept the colleges and the country by mid-century was summed up by a leading New York intellectual:

[27] *Ibid.*, p.47.

[28] I. Bernard Cohen, *Some Early Tools of American Science* (Cambridge, Mass., 1960), quoted on p.11.

[29] Samuel Eliot Morison, *Three Centuries of Harvard, 1636-1936* (Cambridge, Mass., 1937), p.89.

> The school learning is now expelled from the sciences. You'll find nothing of it in the modern astronomy, or in any of the mathematical sciences: the best writers in medicine are ashamed to make use of it. . . .[30]

"By 1740, it is safe to say, the mathematical sciences had established themselves firmly,"[31] although there were only "perhaps a dozen scholars who were able to understand and work with the theory of fluxions and the application of gravitation to celestial bodies."[32] There was no active scientific establishment in the colleges. But a significant change had occurred: there was a new philosophy of education to accommodate new colonial aspirations.

A more complex society demanded educated public servants other than ministers. From the end of the seventeenth century until the early 1740's, the percentage of Harvard and Yale graduates who became ministers fell steadily from 60 to 40 per cent.[33] In the next hundred years the figure would drop to about 20 per cent. With the decrease in religious themes and the elimination of all systematic theology, religion eventually accounted for a very insignificant part of the college program.

The period from 1740 to the Revolution and beyond introduced a new public and institutional attitude toward education. The meaning of the period is too easily missed amidst the recitation of founding dates of new colleges and of new professorships. Often advances forecasted in published announcements were accompanied by little change in classroom practice. No single institution can demonstrate a record of continuous progress in scientific development, as its fortunes were so overwhelmingly dependent upon a single scientist and the funds which at one time or another might be found for his employment. Yet progress there was, as L. F. Snow observed when, in trying to date the introduction of algebra into the permanent curriculum, he concluded: "The exact

[30] Cadwallader Colden, "An Introduction to the Study of Phylosophy Wrote in America for the Use of a Young Gentleman," a manuscript printed in Joseph Leon Blau, *American Philosophic Addresses, 1700-1900* (New York, 1946), p.292.

[31] Theodore Hornberger, *Scientific Thought in the American Colleges, 1638-1800* (Austin, 1945), p.192.

[32] Frederick E. Brasch, "The Newtonian Epoch in the American Colonies (1680-1783)," *Proc. Amer. Antiq. Soc.* **49** (1939): p.323.

[33] Richard Hofstadter, *Academic Freedom in the Age of the College* (New York, 1955), p.192.

year is of little importance, the fact remains that by the middle of the eighteenth century the curriculum of the Colonial College had begun to shape itself along broader lines than those of mere preparation for the ministerial offices."[34]

It is difficult to say how much the proliferation of American colleges beginning with the Great Awakening in 1740 is due to denominationalism, local pride, or real need. There were obvious Old World precedents for each of these causes. But the typically American feature of these colleges was the void in public and secondary education which they helped fill. Not a one had the pretense of becoming an Edinburgh or a Göttingen.

The six new colleges and their legal dates of founding are the University of Pennsylvania (College of Philadelphia), 1740; Princeton (College of New Jersey), 1746; Columbia (King's College), 1754; Brown (College of Rhode Island), 1764; Rutgers (Queen's College), 1766; and Dartmouth College, 1769. Each was under some form of denominational control, but not a single one served as a theological seminary or sent a majority of its graduates into the ministry. Princeton (Presbyterian), the one most obviously formed to fill a religious need, forebade sectarian instruction in the classroom. The charter of Brown (Baptist) required representation of all major Protestant denominations on the governing board. Brown, Columbia, and the University of Pennsylvania, in fact, taught no theology at all.

The oldest colleges—Harvard, Yale, and William and Mary—had already made considerable adjustment to eighteenth-century intellectual reality by midcentury. A completely new set of educational values entered with the rise of Columbia and the University of Pennsylvania. William Smith, educated in the Scottish Enlightenment at Aberdeen, helped to found Columbia and became first provost of Pennsylvania. Although denying neither the value nor the necessity of classical education and the seven liberal arts, Smith placed emphasis on mathematics, natural science, and the English language. Here, for the first time, in Smith's proposals, is provision for instruction in natural history, agriculture, botany, and chemistry. Smith's ideal was to prepare the

Great Awaken-ing 1740

6 new colleges

Wm Smith

[34] Snow, *op.cit.,* p.52.

student to live an enlightened life rather than to prepare him for the ministry or magistracy.

Despite the new receptiveness to scientific disciplines, Edward Potts Cheyney has pointed out that in practice much of the instruction in modern science "can have been little more than a series of definitions."[35] This is true not just of Smith's program, carefully executed until the Revolution, but of programs at all colleges before the second quarter of the nineteenth century.

There were many factors hindering rapid transformation of the American college, still no better than a high school, into a European type center of learning. There was not a continuous supply of adequately educated teachers of science. No college escaped the experiences of having science professorships vacant or lectures advertised but never delivered. There was still no adequate system of pre-college training; and the colleges themselves, "filled with children,"[36] labored under severe financial strain.

Still, the acquisition of "philosophical apparatus" was considered a virtue as soon as interest in experimental science washed ashore. Harvard, Yale, and William and Mary had considerable and varied demonstration pieces by the 1730's, and no school after midcentury attempted to instruct without instruments. Yet the fact that so much of the collections remains intact today, particularly the delicate electrical curiosities, indicates that they were not constantly handled by students, if at all. Teaching science consisted "either of daily recitation from textbooks, or of lecturing with the help of whatever apparatus the college could afford."[37]

The Revolution caused only a temporary setback to the realization of Enlightenment ideals. Although the histories of all nine colonial colleges save Dartmouth are full of patriotic reminders of how their campuses were occupied by troops of one side or the other and how student attendance declined, there was no post-bellum collapse into archaic thinking. The ideal of the utility of science had not been lost, and interruptions of activities such as those of the American Philosophical

[35] Edward Potts Cheyney, *History of the University of Pennsylvania, 1740-1940* (Philadelphia, 1940), p.83.

[36] Samuel Miller, *A Brief Retrospect of the Eighteenth Century* (New York, 1803) 2: p.274.

[37] Hornberger, *Scientific Thought*, p.35.

Society did not cause any deterioration in the mathematical or astronomical program when full college activity resumed in the 1780's. Individual colleges made new financial and political arrangements with the new state governments, but there was no change in philosophy or instruction.[38]

After the Revolution there was another wave of new colleges so that, by 1800, twenty-six were in operation. By 1828, the number had risen to fifty. Most of the newer colleges were located in the Northeast although there were about a dozen colleges in the South and a few across the Alleghenies. Most, even when they had closer associations with the states, were in some way under denominational influence. All of the new colleges displayed the distinctly American system of government by an outside board of non-teachers. All began as small, isolated schools with no provision for professional study. All were in serious financial difficulties well into the nineteenth century, if they survived at all.

new colleges

In 1800 not a single college had a teaching staff capable of rendering high-level instruction in science. We may ask, how did the college of 1800 differ from that of 1750? A contemporary, Samuel Miller, wrote in 1803,

> In America the decline of classic literature is especially remarkable and prevalent. Many of our colleges require in their students but a superficial acquaintance with the Latin language; and with respect to Greek, are contented with a smattering which scarcely deserves the name of knowledge. . . The causes of this revolution are various. Since the commencement of the eighteenth century, the physical sciences have been gradually extending their bounds, demanding more attention, and acquiring greater ascendancy. Another cause which has doubtless contributed to the effect in question, the inconceivable enlargement and activity which the past age exhibited. New objects of profit and pleasure have arisen, and engaged the public mind; new fields of labour and adventure have been thrown open; and of course, in calculating an education for active life, the refinements of ancient literature began to receive a smaller share of regard . . . . It must be admitted that this manifest decline of classic literature has been attended with some advantages. In consequence of discarding dead languages, as the ordinary me-

[38] Russell Blaine Nye's *The Cultural Life of the New Nation, 1776-1830* (New York, 1960), chap. 4, "The Organization of Scientific Knowledge," pp.70-96, summarizes society's (as opposed to the college's) scientific concerns after the Revolution.

> dium of philosophical publication, such writings have become accessible and popular: the student has some time left for becoming acquainted with his vernacular tongue; the attention of the learned is more directed to moral and physical science.[39]

Undoubtedly, the "enlargement of the sphere of enterprise" included medical schools and societies, natural history clubs, literary and scientific societies, agricultural societies, and societies for the promotion of arts and manufactures—all legacies of the last half of the eighteenth century.

Early American colleges were slowly changing institutions. One Catholic historian has devoted an entire book to the demonstration that scholastic thinking was still important in the colonial college and lasted well into the nineteenth century.[40] An inspection of the form and content of these at the end of the century certainly shows that scholastic techniques were not yet lost. The hope of being able to gather all knowledge into a few volumes persisted.

Science was present and few doubted that its importance would increase. I. B. Cohen reports he has "looked carefully through the records of science in our Colonial colleges for any evidence of an organized anti-scientific spirit and . . . found none."[41] But as yet there had been no evidence that colleges could fill any of the utilitarian needs of the surrounding community. Isolated, excellent professors dotted the eighteenth-century landscape, but no college was able to capitalize on their work and create a center of scientific inquiry or even of scientific education. Or, as Miller lamented,

> in some respects, and particularly with respect to the patient, laborious and thorough investigation of the various objects of knowledge; the depth of erudition; the discipline and subordination of academic establishments; and the general moral influence of literary and scientific acquirement, the last age cannot with propriety boast of much progress.[42]

But even if science had not yet triumphed, neither did theology reign. At the beginning of the nineteenth century, there

---

[39] Miller, *Brief Retrospect* 2: pp.37-39.

[40] James J. Walsh, *Education of the Founding Fathers of the Republic* (New York, 1935).

[41] I. Bernard Cohen, "Some Reflections on the State of Science in America during the nineteenth century," *Proc. Nat. Acad. Sci.* 45 (1959): p.669.

[42] Miller, *op.cit.* 2: p.302.

existed neither an excessive devotion to the classics nor a desire to teach theology. The General Synod of the Dutch Reformed Church decided in 1784 that its College, Rutgers, need not even elect a professor of theology. It established instead the first theological seminary in America in New York. Soon theological seminaries of the other denominations followed. By the time that Yale abolished its faculty orthodoxy tests in 1831, the secularization of the northeastern college was fairly complete. And classics, too, occupied an increasingly diminished role in the curriculum.

II. The Revolution of the 1820's

THE thirty-year period between the two wars, 1783–1815, found American interest in science growing, but at a lesser pace than at the intellectual centers of Europe. The number of contributors to science was still so small that the whole state of particular scientific disciplines was determined by a few individuals who practiced them. Not until 1820, according to an historian of chemistry in America, did the growth of that science become "definite and even and not so obviously dependent upon the life and death of any particular man."[1]

America, still overwhelmingly rural, was feeling less of that pressure for change which the Industrial Revolution was effecting in European society. Before the Revolution, cultural fertilization had been unavoidable: colonial America was an appendage of the European intellect and eager to reflect the ideas of the Old World. But after the war, anti-English feeling and later anti-Jacobin feeling formed effective barriers against the swift invasion of European interests and institutions into America.

Within the thirty years in question, European knowledge rendered obsolete much of what was contained in the corpus of "natural philosophy" in America. A new planet and numerous new elements were discovered, the effects of infrared and ultraviolet (actinic) radiation exposed, the science of geology born, the power of steam efficiently harnessed, and the first fruits of scientific agriculture harvested. With literary channels to Europe certainly not closed, it is no wonder that "somewhere around 1815, Americans became interested in the pursuit of science to a greater extent than ever before."[2] Even

[1] Robert Siegfried, "A Study of Chemical Research Publication from the U.S. before 1880" (unpublished doctoral dissertation, University of Wisconsin, 1952), p.59.

[2] George H. Daniels, *American Science in the Age of Jackson* (New York, 1968), p.3. See also John C. Greene, "American Science Comes of Age, 1780–1820," *Jour. Amer. History* 55(1968): pp.22–41.

the *North American Review,* one of the new general journals influencing popular opinion, could promise would-be contributors of articles on science that their work would be "extensively circulated" and "generally read in the Country."[3] A scientist could applaud the fact that "a great number of young and able observers or writers begin to appear everywhere."[4] The colleges, however, had not grasped the implications for education of these developments.

Science in the colleges was not, to be sure, defunct. The schools continued to offer at least some mathematics and natural philosophy as they had before the Revolution. But while young American college students continued to be amused by static electrical devices and stories about Benjamin Franklin, a national hero, Europeans had entered the new era of current electricity. As late as 1816, the Harvard scientist complained that the College did not have a battery sufficient to charge a Leyden jar and hence demonstrate the equivalence between current electricity and that produced by a static generator. Without such equipment for verification, he lamented, "we are obliged to rely upon authority."[5]

Improvements had, however, been made in the execution of the college's enlightenment philosophy; certainly, the college of 1816 was different from the college of 1790 in its scientific efforts. According to Benjamin Silliman, a freshman at Yale College in 1792, "Chemistry was, then, hardly known in the Institution, even by name . . . there were a few articles of natural history, in the miscellaneous museum, but they had little or no reference to science. . . . As to philosophical apparatus, there was no more than was included in a common (student) room."[6] Yet by 1802 Silliman, like scattered others who had entertained no thought of teaching science, would be requested to prepare himself to deliver scientific lectures.

During the first few decades of the nineteenth century, all colleges offered some scientific training. "In all . . . [recorded

[3] MS, Edward Everett to Parker Cleaveland, April 29, 1820, University of Virginia Library.

[4] C. S. Rafinesque, "Survey of the Progress and Actual State of Natural Science in the United States of America, from the beginning of this century to the present time," *Amer. Monthly Magazine and Critical Rev.* 2(1817): p.81.

[5] MS, John Farrar to President [Kirkland], Dec. 27, 1816, Harvard College Archives.

[6] MS, Address to Alumni, July 22, 1842, Yale University Library, Stokes Autograph Collection.

one observer in 1817] natural philosophy is taught; in some of them chemistry; in a few natural history.[7] For a short time between 1796 and 1806, Princeton even admitted special non-degree students to its scientific lectures—a tradition established more firmly at Union College with its adoption of the experimental partial course program designed by science-minded President Eliphalet Nott. Men such as James Dean at the University of Vermont, Parker Cleaveland at Bowdoin, and Chester Dewey at Williams were diverted from traditional ministerial careers to fill the temporary voids in their institutions' academic programs. By 1819, the *North American Review* could applaud the "method recently adopted in several colleges of selecting Professors, . . . and giving them opportunity to make preparation for their particular duties before entering upon them."[8] A less sanguine, but still optimistic observer assured that "instruction will flourish with more rapidity when they shall be selected, in all instances, among the most worthy and learned candidates," when those candidates should become available.[9]

For every flattering remark, however, that an optimist might make, there was indication that the level of instruction and attention to science was low. Well into the century, college classes paid little attention to even those poor textbooks that existed.[10] To no one's great consternation, colleges neglected science in those years either because they could not find professors to assume its instruction or because the men appointed were occupied with too many other duties, such as the teaching of mathematics or even Latin and Greek.[11] Often colleges tried to save money by having part-time instructors, frequently local medical doctors, deliver lectures on the rudiments of physics and chemistry, these lectures being sometimes not even included in the comprehensive tuition fee.

Most of the men who taught these outdated children's exercises worked in unfavorable environments. Unlike other professors, they were often appointed for one year at a time. Scien-

[7] Rafinesque, *op. cit.*, p.84.

[8] *North American Rev.* **9**(Sept., 1819): p.250.

[9] Rafinesque, *op. cit.*, p.84.

[10] A detailed description of science and student accomplishments at early nineteenth-century Harvard is provided in John Gorham Palfrey, *Notice of Professor Farrar* (Boston, 1853), p.9, *passim.*

[11] See *North American Rev.* **9**(Sept. 1819): p.249 for remarks on professors.

tific apparatus was expensive and difficult to obtain in this country. Almost without exception, the colleges suffered from critical financial shortages; and science was always the first area of instruction to be discouraged by the lack of funds. A new source of spirit and wealth was required. At Princeton, John Maclean, who was trained in Scotland, and probably the first to teach Lavoisier's new chemistry in America, wrote to his friend and fellow chemistry teacher at Yale, Benjamin Silliman: "In regard to chemistry you ought to be thankful that you are only liable to 'growing rusty'—as far as me, owing in great measure to certain arrangements in our college, I have paid no attention to it [chemistry] for the past eighteen months."[12]

In short, although not impervious to the intellectual demands of the newer science, the older colleges—and the newer ones established in their pattern—delivered the same courses and served the same function after the War of 1812 as they had on the eve of the Revolution. The core of the curriculum was still oriented toward ancient languages and culture. Faculties remained powerless to suggest or implement changes since they had never coalesced into homogeneous groups for the sake of bargaining with the all-powerful trustees. There were no precedents for suggesting curricular reform or self-evaluation. The handful of gentlemen who "heard the recitations" were collectively known as the "immediate government of the college," and there is reason to believe that discipline occupied as much of their time and effort as actual instruction. The failure of eighteenth-century paternalism and curricular rigidity is suggested both by the formation of student literary societies and by the uprisings which clearly mark the first quarter of the nineteenth century as the most disruptive in American college history. Indeed, Yale president Timothy Dwight's last wish for the college, made on his deathbed in 1817, was that "discipline may be preserved."[13]

Forces besides that exerted by science were operating to

[12] MS, John Maclean to Benjamin Silliman, Feb. 10, 1810, Princeton University Library.

[13] Quoted in *Remarks on the Present Situation of Yale College for the Consideration of its Friends and Patrons* [no author], printed January, 1823, p.16. For the view that discipline was, at least, no more of a problem in American colleges than in the European universities, see "On the means of education and the state of learning in the United States of America," *North American Rev.* **9**(1819): p.245.

eliminate these unfavorable conditions. Many students and professors alike were interested in the emerging social science theories, in modern languages, and modern civilization. But the greatest reason for the college's introspection would be provided by the changing pattern of secondary education. The proliferation of academies since the Revolution—15 in Massachusetts by 1797, 55 in Pennsylvania alone between 1784 and 1819, and almost a thousand in the country by 1830—had given the colleges something with which to compare themselves.[14] Of the 149 new subjects which were to appear in the "academical" programs of New York State between 1781 and 1870, one-half came in the single three-year period preceding 1828.[15] These new subjects included algebra, astronomy, botany, and chemistry. Among the consequences of this broadening of the preparatory curriculum, as many were soon to realize, was that a college student entered college with some science, however poorly taught, as part of his background. The more advanced age at which students entered college was another important result of the establishment of the American academy. Gone were the eighteenth-century days when twelve-year-old boys could occasionally frequent collegiate halls: in the 1820's the seventeen-year-old graduate was to provide the colleges "with a large part of their students."[16] If academies were providing classical and scientific instruction as sophisticated as that offered by many colleges, what was the colleges' *raison d'être?* The answer, obvious to many, was that "the colleges succeeding the academies, should accommodate their course of study to the advances already made by the pupils of the academy."[17] But how might the college take advantage of the students' increasingly better preparation to guarantee interest in still higher education?

These problems of the college were intensified and more

---

[14] For Massachusetts, see Claude Moore Fuess, *Amherst: The Story of a New England College* (Boston, 1935), p.23. For Pennsylvania, *Register of Pennsylvania* 2, no.20, Philadelphia, Nov. 29, 1828, p.312. For total, Ellwood P. Cubberley, *Public Education in the United States* (New York, 1919), p.185; Cubberley cites the origin of commonly used academy statistics.

[15] Cubberley, *Public Education,* p.187.

[16] William Fowler, *A Lecture on the Influence of Academies and High Schools on Common Schools, delivered in Boston before the American Institute of Instruction, August 1831* (Boston, 1831), p.1.

[17] Charles Fenton Mercer, *A Discourse on Popular Education* (Princeton, 1826), p.74.

widely discussed than ever before in the great wave of periodical and pamphlet literature of the nineteenth century. In 1825 it was remarked that "no demand has been more constant, more extensive, or more earnest, than the demand in this country, for an improved state of education."[18] With the physical sciences being cultivated in the first few decades of the century "with an energy and success, surpassing all former example," it seemed almost inevitable that science should play a major role in discussions about curricular reform.[19]

Individual colleges sensed that their existences were threatened in a changing social environment; for the failure of student enrollment to keep up with population or popular enthusiasm warned them that their academic programs were unsuited to the times. However, as local institutions, colleges had little experience in cooperating with their neighbors; and the problems which affected all had not yet been clearly defined.

Thus, with only a vague sense of the changes which might occur in the next decade, representatives from Vermont, Harvard, Union, Bowdoin, Middlebury, and Yale gathered in New Haven on September 18, 1818 (less than a month, incidentally, after the first volume of the *American Journal of Science and Arts* appeared) for a "special meeting of the instructors of Colleges in the Northern section of the United States." Although the stated purpose of the meeting was "the general interest of letters," its specific nature was soon apparent. John Kirkland, the president of Harvard, and Jeremiah Day, the president of Yale, formed a committee "to propose measures to effect a greater uniformity in the requisite attainments for admission into College, and in the books to be used previously and subsequently to admission." This association of professors, enlarged by the addition of members from Brown, Dartmouth, and other colleges, reconvened on May 24, 1819, in Boston's Court House as the "first annual meeting of the Collegiate Convention, consisting of gentlemen connected with the Colleges and Universities in that section of the United States lying North of the Delaware." Again the topics

---

[18] George Ticknor, *Remarks on Changes Lately Proposed or Adopted in Harvard University* (Boston, 1825), p.3.

[19] John Griscom, *A Discourse on the Importance of Character and Education in the United States* (New York, 1823), p.8.

of discussion were "preparatory studies, admission [requirements], and studies subsequent to admission."[20]

These meetings were a prelude for a new pattern of institutional cooperation and competition. They came at the same time that the Supreme Court decision in the Dartmouth College Case (March 2, 1819) declared that colleges were organized by private contract independent of the state, thus opening an era of college founding in which all colleges had to vie for survival.[21] The overwhelming majority of the colleges published their first descriptive catalogs shortly thereafter. Once collected, the catalogs, along with circular letters of inquiry, provided faculties with strong support for the requests they made to their trustees. In the twenties, there emerged a nexus of competition among the Northeast colleges which provided that a successful innovation in one would be quickly repeated at the other schools and guaranteed as well that colleges would be afraid of too radical innovation. The result was that, despite denominational, geographical, and social varieties, the colleges presented such similar curricula that they were in some ways almost indistinguishable.

Tense excitement over education in the 1820's ended a half-century of relative complacency. Critics of higher learning could be found around every corner. To some, the problem with the colleges seemed to be their failure to reproduce models of continental scholarship; while to many others, the problems were those of organization and discipline. The most common complaint, which summarized a melange of frustrations, was that the college did not partake of the "spirit of the age": Why, for instance, did they fail to provide vocational training for the manual classes?

Thus, with the colleges exposed to view and with the charges against them many and varied, it is not surprising that professional educators and amateurs alike seized upon a variety of circumstances to develop a philosophy or at least a blueprint for college education. Published results, in the form of reports from faculties, presidents, and trustee-committees, represent an evaluative procedure which, if not always effec-

[20] MS, Levi Hedge to William Harris, March 20, 1820, Columbia College papers, Columbia University, describes these meetings.

[21] Donald G. Tewksbury, *The Founding of American Colleges and Universities before the Civil War* (New York, 1932) is the standard work on the growth and death of colleges before the Civil War.

tive, was certainly novel. Traditional historiography errs, then, in its point of view that the post-1825 deliberations were unsuccessful or at best premature, simply because they failed to effect the establishment of a modern university system.[22]

Quite naturally, Harvard, the oldest and wealthiest of the American colleges, was the first to serve as a forum for new educational views. Following the undergraduate rebellion of 1823 and the subsequent arguments over institutional polity, faculty member George Ticknor decided to state his views on the whole of the college predicament. Ticknor, himself a Dartmough graduate (1807), was one of the first Americans to take advantage of German university training and became Smith Professor of Modern Languages at Harvard in 1819. He thought that "exaggerated fears and statements" had arisen from the uncertainty surrounding the "indistinct persuasion [that the college] did not keep pace with the spirit of the times,"[23] and he proposed to clarify the nature of the collegiate evil.

First in Ticknor's list of complaints was the fact that over twenty weeks in the year were wasted on vacations. Next under attack was the system of student parroting at recitations and the grouping of students of all abilities in the same class, "giving to the most industrious and intelligent no more and no other lessons, than to the most dull and idle." The result of the recitation system, according to Ticknor, was that the students "learnt a given *book,* rather than a given subject"; the lectures he decried as a "waste of time both of the hearers and instructors." Indeed, "the great principle of *teaching* was lost sight of."[24]

By the end of Ticknor's forty-eight-page essay, each part of the trilogy which then defined the "college" had been blasted —organization, discipline, and instruction. The suggestions for reform ranged from simply shortening vacations to invoking a university system after the German model—a suggestion

[22] Richard J. Storr, in *The Beginnings of Graduate Education in America* (Chicago, 1953), evaluates the 1820's in terms of the graduate school and fails to recognize the achievements of its somewhat undirected revolution. A more sympathetic view may be found in Frederick Rudolph, *The American College and University* (New York, 1962), where an "unseen revolution" (p.136) is defined as resulting from these proposals.

[23] George Ticknor, *Remarks on Changes Lately Proposed or Adopted in Harvard University* (Boston, 1825), p.4.

[24] *Ibid.*, p.5.

which bore little fruit at the time. Some lesser proposals were actually implemented—the division of instruction into distinct departments, the grouping of students according to ability, and the theory that "students shall, to a certain degree, have a choice in the studies they are to pursue."[25]

But the suggestion which appeared first on Ticknor's list of proposals spoke directly to the complaints coming from outside academe: Why should Harvard "confine its instructions to a strictly marked course, the end of which is limited by a merely formal and unmeaning certificate"? Ticknor thought that a large college should "open its doors to all." For example, "Harvard College has abundant resources to render unnecessary the establishment of many public and private institutions, like the agricultural schools, the law schools, and the other establishments for special purposes, which like the admirable and flourishing institution of Gardiner, in Maine, are already beginning or begun among us also." Ticknor, like others, recognized the increasing failure of the college to provide "a liberal education for many persons in all classes of the community,"[26] and Harvard responded to his criticism. Rejecting Ticknor's most radical proposal to make college attractive to everyone, Harvard did, nevertheless, immediately make "considerable changes" in its science programs: late in 1825 chemistry was finally taught regularly to undergraduates under the auspices of a "professed teacher, devoted to the subject," and preparation was made for similar instruction in natural history.[27]

The "inadequacy of the prevailing systems of classical education in this country to meet the wants and demands of an enlightened public" also aroused the faculty of newly created Amherst. Led by Jacob Abbott, professor of mathematics and natural philosophy, who had studied with Parker Cleaveland at Bowdoin in the class of 1820, the faculty presented a report to the trustees on the state of the college (August 21, 1826). In response, the trustees appointed their own committee to inspect the charges and requested the faculty to be more specific

[25] *Ibid.*, p.38.

[26] *Ibid.*, pp.35-36. See Rudolph, *American College and University*, pp.118-121, for a discussion of the reception and results of Ticknor's proposals at Harvard.

[27] MS, John Farrar to [President and Fellows], December 26, 1825, College Papers (XI), Harvard College Archives. Also, Farrar to Overseers, Jan. 10, 1826.

in their criticisms. The results were published early in 1827 following a special meeting on December 6.[28]

The Amherst proposals covered some of the same ground that Ticknor had—public dissatisfaction with colleges as they existed, the need to amplify instruction in modern language and literature, and the inability to service "merchants and farmers." They also raised the issue of the position of the classical languages in the curriculum, arriving at the general admission that there were legitimate ways in which some students could avoid their continued study; and they strongly advocated a "department of theoretical mechanics." A new parallel course designed to bypass some of the older classical studies would emphasize "Chemistry and other kindred branches of Physical Sciences by showing their application to the more useful arts and trades, familiar lectures upon curious and labour saving machines, Natural History," and other recently evolved disciplines. The Amherst faculty, as many others, was imbued with the idea of progress; and in assessing the competing claims of moderates, liberals, and even reactionaries in educational reform, it found one fact with which all of them "entirely agreed":

> These institutions do not at present, afford all the facilities which they want, for the liberal education of their sons; and we are convinced, that if the Colleges cannot so modify their systems, as to meet the public demand, or if they do not choose to do it, other seminaries, equal in rank and of surpassing popularity, will spring up by their side.[29]

The University of Vermont, which had been on the "verge of extinction" in 1821, came to grips with the vexing problem of public demand in 1826. The University found new strength in the appointments of science professor George W. Benedict

Amherst

Chemistry and its appl. natl hist.

Vermont George Benedict

---

[28] *The Substance of Two Reports of the Faculty of Amherst College to the Board of Trustees, with the doings of the Board Thereon* (Amherst, 1827). The reports are summarized in *North American Rev.* 15, (n.s. April, 1827): pp.485-487. According to the Faculty Records (MS, Oct. 16, 1826, Amherst College Library), Ebenezer Snell, then a tutor, aided Abbott in examining "the several plans for two parallel courses." A rumor conveyed to Professor George W. Benedict of the University of Vermont suggests that Edward Hitchcock authored some of the report (MS, Benedict to Joseph Torrey, April 7, 1829, University of Vermont Archives). It would then appear that at Amherst the entire reform movement was headed by scientists.

[29] *Substance of the Proposals* . . . , p.7.

in 1825 and President James Marsh in 1826. As soon as Marsh arrived in Burlington, he enlisted the aid of the faculty in a scheme to reorganize the college, permitting students to pursue courses in any or all of four departments of knowledge—English Literature; Languages; Mathematics and Physics; Political, Moral, and Intellectual Philosophy. The classical languages would only be insisted upon in the case of those students who were candidates for the degree, which itself was to be earned after a certain amount of advancement at the student's individual pace, rather than after four years of physical presence at the University. The most novel provision was a liberalization of the role of the student as citizen on campus and the elimination of traditional forms of punishment.[30]

The plan, adopted in the fall of 1827, proved effective. After the appearance of the Amherst and Yale reports, Benedict decided to make known the Vermont plan. "The occasion [Yale's Report in *Silliman's Journal* and the Amherst proposals in the *North American Review*] seemed a very fine one for us to come forward with some account of our own. We thought at first to publish it in the same work [AJS], but afterward judged best to put it in pamphlet."[31]

Much like Amherst and Harvard, Vermont called for the college to "be opened to all who need any assistance which it can afford."[32] But what was no clearer to the visionaries of Vermont than to those at Harvard or Amherst was that the colleges might at least temporarily service the changed requirements of a changed society and economy by suitable modification of the college's central curriculum, without distorting the ancient ideal of the liberal arts education. It is in this context that we must understand the origin, reception, and execution of one of the most important evaluative surveys in American education—the Yale Report of 1828.

Yale, the largest American college in the 1820's, had more students than Harvard but many fewer resources. Boasting the

---

[30] Julian Ira Lindsay, *Tradition Looks Forward: The University of Vermont, A History, 1791-1909* (Burlington, 1954), p. 130, *passim*.

[31] MS, George W. Benedict to Joseph Torrey, April 7, 1829, University of Vermont Archives. *An Exposition of the System of Instruction and Discipline pursued in the University of Vermont* by the Faculty [G. W. Benedict] (2nd ed., Burlington, 1831). Lindsay, *Tradition* . . . attributes authorship to Benedict, p.139.

[32] *An Exposition* . . . , p.5.

most geographically diverse student body, it was preparing itself to become the most influential college in the coming age.[33] Soon after the appointment of President Jeremiah Day in April, 1817, however, Yale demonstrated concern for its survival. By 1830, it would be almost bankrupt. Williams, Union, Middlebury, and Hamilton had been diminishing its influence in parts of New York, and New England since the beginning of the century; and in 1823, a pamphlet appeared intimating the depth of Yale's despair.[34]

In this pamphlet, primarily a plea for individual benefactors to keep the College alive, hopes were expressed that scientific instruction and the library would increase. There was a suggestion that "changes in the old system [of classifying students] are contemplated." But the instructional problem most clear to the author was the necessity of a "division of labour" among the professors. No longer could the college call upon one man to teach more than one department of knowledge.[35]

Yale knew of the discussions at Amherst, Harvard, and the University of Virginia, which had become a center of elective reform in 1825. It also knew the public complaints of the period and had experienced problems within its own walls. Thus, in September, 1827, when the Corporation requested only an inquiry into the status of the "dead languages," the faculty responded with a report in which all the major educational suggestions of the time are carefully dissected.[36]

The report falls into two parts: the longer, written by President Day, summarizing the "plan of education in the college" and the shorter, written by classicist James L. Kingsley, dealing more specifically with the role of the classical languages. The theme which pervades the entire report is not

[33] Complete statistical tables for all colleges including number of alumni, number of students from each state at each college, and even condensed course of instruction may be found in the *Quarterly Register and Journal of the American Education Society* (title varies), (Andover). These conclusions are drawn from Vol.1 (1829), pp.224-226. Yale was the controlling force of the American Education Society.

[34] *Remarks on the Present Situation of Yale College for the Consideration of its Friends and Patrons* [no author, possibly President Day] (printed January, 1823).

[35] *Ibid.*, p.6.

[36] The report was published as *Report of the Course of Instruction in Yale College; by a Committee of the Corporation and the Academical Faculty* (New Haven, 1828) and reprinted in its entirety in 1830. It was widely known through the *Amer. Jour. Science and Arts* 15 (1829): pp.197-351 as "Original Papers in Relation to a Course of Liberal Education." Page references are to the original published version, 1828.

the necessity of Latin and Greek, the denial of Ticknor's radical reforms, or a passive commitment to procedures of the past, but rather a definition of and reliance upon collegiate education. It stood, both then and today, as a grand answer to all the inchoate suggestions that colleges did not fit the scheme of Jacksonian democracy because they failed to teach all the skills by which some men might achieve their measure of financial success.[37]

The Report was comprehensive, open-minded, and liberal, even if its purposes were not those of the "university" or the "manual skills" factions. In fact, it expressed great surprise at the popular "suggestion that our system is unalterable." Near the beginning and in numerous places throughout the Report, the faculty voiced its willingness to accept changes:

> We are decidedly of the opinion, that our present plan of education admits of improvement. We are aware that the system is

[37] Since my reading of the Yale Report is crucial to my understanding of the educational climate in which scientific instruction was to be advanced, I think it necessary to provide references to alternative interpretations.

R. Freeman Butts, *The College Charts Its Course* (New York, 1939), pp.117-125, discusses the Report in a chapter entitled "The College Controversies Take Shape: Conservative Reaction," emphasizing to great extent Yale's reluctance to adopt an elective system.

Richard J. Storr, *The Beginnings of Graduate Education in America* (Chicago, 1953), *passim,* is similarly impressed by the Report's ability to curb university growth. Storr does, however, warn against the conclusion that the college course was unchanging, even though these changes were not his concern (p.5).

Richard Hofstadter and Wilson Smith, *American Higher Education: A Documentary History* (Chicago, 1961) 1:pp.275-291, have condensed the Report in such a way as to emphasize its negative aspects. However, Hofstadter and C. Dewitt Hardy, in another volume, *The Development and Scope of Higher Education in the United States* (New York, 1952), pp.15-17, warn of the "temptation to overrate its educational value," while pointing out the "elements of moderation and wisdom in the report" (footnote, p.16).

Frederick Rudolph, *The American College and University* (New York, 1962), pp.130-135, after dwelling upon the failure of the elective and university principles to take hold and upon the "aristocratic purpose" (p.134) of the Report, also introduced a discussion of the beneficial effects emanating from the 1828 tradition.

George P. Schmidt, *The Liberal Arts College: A Chapter in American Cultural History* (New Brunswick, N.J., 1957), pp.55-58, leans toward conservative interpretation but adds: "The cultural values of science, it is only fair to say, were also vigorously defended."

Walter P. Metzger, *Academic Freedom in the Age of the University* (New York, 1955) thinks he has found in the Report "a consistent spirit and argument: the preceptive importance of religion" (p.5), despite the fact that religion does not enter the discussion in even the most obscure way; in fact, the work "religion" does not even appear. It would seem that there is still room for further interpretation of the 1828 tradition.

> imperfect; and we cherish the hope, that some of its defects may, from time to time be made with advantage to meet the varying demands of the community to accommodate the course of instruction to the rapid advance of the country, in population, in refinement, and opulence. We have no doubt that important improvements may be suggested, at attentive observations of the literary institutions in Europe; and by the earnest spirit of inquiry which is now so prevalent, on the subject of education.[38]

need for improvement

As examples of change within the previous thirty years, they could point to the introduction of "whole [new] sciences —chemistry, mineralogy, geology, political economy."[39] In the same vein, they could complain of the folly of popular expositors who suggested that from 1714 to 1828 "only 'slight alterations' have been made in the system of education in this college."[40] But the faculty, though wholly receptive to "improvements . . . without hazarding the loss of what has been already attained," felt constrained to operate within a collegiate framework and hence to ask, "What then is the appropriate object of a college?"[41]

Their answer to this question was that the "object is to LAY THE FOUNDATION OF A SUPERIOR EDUCATION."[42] This foundation consisted of the proper "discipline" and "furniture" of the mind. To succeed, it was necessary that "*all* the important mental faculties be brought into exercise."[43] Although "faculty psychology" as a paradigm of intellectual development may today no longer be an adequate description of the learning apparatus, as a paradigm for curriculum planning it was then still meaningful. It emphasized the variety of courses needed to sharpen all the students' powers; and it insisted upon "throw[ing] the student upon the resources of his own mind."[44]

faculty psych.

---

[38] *Report,* p.5

[39] *Ibid.,* p.6.

[40] *Ibid.,* p.45.

[41] *Ibid.,* pp.6, 7.

[42] *Ibid.,* p.6.

[43] *Ibid.,* p.7.

[44] *Ibid.,* p.8. Laurence R. Veysey, *The Emergence of the American University* (Chicago, 1965), pp.22–25, gives a description of the nineteenth-century conception of the psychology of the mental faculties. In this context, Veysey defines faculty as "specific capability or potential talent" (p.22). Walter B. Kolesnick, *Mental Discipline in Modern Education* (Madison, 1958), Chapter V, pp.89-112, gives the history

In practice, it meant that "languages alone, or mathematics alone, or natural or political sciences alone" could not properly invigorate all the faculties of reasoning, demonstration, memory, and invention.[45] If ancient languages could not be set aside without destroying the system, neither could "the various branches of the mathematics, or of history and antiquities, or of rhetoric and oratory, or natural philosophy, or astronomy, or chemistry, or mineralogy, or geology, or political economy, or mental and moral philosophy."[46] Although perhaps it sounds as if the choice of studies was considered unalterable, we must recall that the whole report grew out of an earnest desire to evaluate the place of the classical languages in the curriculum and that the study did propose that "as knowledge varies, education should vary with it."[47]

Having defined the nature of education, the Report went on to describe what it should not seek to become. Briefly, it ought not to incorporate "every thing which the student will ever have occasion to learn,"[48] "professional studies" of any kind, nor the "minute detailes of mercantile, mechanical, or agricultural concerns."[49] In essence, a college was not to be all things to all people:

> The public are undoubtedly right, in demanding that there should be appropriate courses of education, accessible to all classes of youth. And we rejoice at the prospect of ample provision for this purpose, in the improvement of our academies, and the establishment of commercial high schools, gymnasia, lycea, agricultural seminaries, &c. But do the public insist, that every college shall become a high school, gymnasium, lyceum, and academy? Why should we interfere with these valuable institutions? Why wish to take their business out of their hands?[50]

---

of the concept of mental faculty and then suggests that the Yale Report shaped its meaning for the rest of the century (p.11).

[45] *Report,* p.9.

[46] *Ibid.,* p.19. The reader may use this list or any post-1825 catalog to satisfy himself that religion, though part of the college social structure, played no role in the curriculum.

[47] *Ibid.,* p.31.

[48] *Ibid.,* p.14.

[49] *Ibid.,* p.15.

[50] *Ibid.,* p.24.

Yale did not see in its immediate future any move toward university status. However, the Report declared that, if Yale ever became like the University of London with "higher and inferior courses, its scientific and practical departments, its professional, mercantile, and mechanical institutions . . . our present under-graduate course ought still to constitute one distinct branch of the complicated system of arrangement."[51] The most "conservative" aspect of the Report was its insistence on conserving the idea of a college. Whatever criticisms the Report has borne and will continue to bear must always sound like the cries, heard in every period, that intellectual pursuits are "out of spirit with the age."[52]

One section of the Yale Report concluded that the "changes in the country, during the last century, have not been greater than the changes in the college."[52] And at no time since 1740 had the changes been more fundamental than in the decade 1818-1828. Faculties were now constantly concerned with "how the course of education in the college[s] might be improved,"[53] and had established lines of communication with the trustees. Alumni groups were established. An American textbook industry stood ready to serve a more sophisticated curriculum. Students were better prepared. Change was accepted as a fact of collegiate life. Following the Yale guideline—"As knowledge varies, education should vary with it"—this group of colleges would survive, grow, and provide college students with the largest doses of science they would ever all be required to take.[54]

The vicissitudes of reform were not confined to Amherst, Harvard, Vermont, and Yale. Similar problems of organization, discipline, and instruction beset the other schools, all of whom responded by expanding their liberal arts studies and including more science.

When Samuel L. Mitchill surveyed Columbia College in 1794, he was able to report that the professor of natural philosophy and astronomy was actually delivering his lectures. Only three years earlier, a student, with a dozen pages of notes for

[51] *Ibid.*, p.25
[52] *Ibid.*, p.46.
[53] *Ibid.*, p.45.
[54] *Ibid.*, p.31.

the year, had remarked that these lectures were "commenced . . . and never ended"—an instance indicative of the state of scientific instruction at the turn of the century.[55] Columbia, however, located in an increasingly cosmopolitan New York, came more easily under public scrutiny than most colleges. A committee of the trustees made an early study of the school in 1810 and, in its defense of the American collegiate arrangement, suggested a program that offered some mathematics or natural philosophy every year. In 1813 Columbia's joint professorship for those subjects was filled by the talented Robert Adrain, who wanted the college to progress faster than the trustees were willing. The following year, he even proposed that Columbia build an observatory.[56]

Adrain's persistence and habit of covering more mathematics than the students could follow prompted the trustees to make a study of his proper functions. They concluded,

> The duties which the Professor of Mathematics and Natural Philosophy is at present required to perform are too numerous and various to be discharged by an individual, in a manner consistent with the interests and character of the College, and the improvement of the students. A division of that professorship would give a greater scope to the useful exercise of those mathematical talents for which the gentleman who fills that station is so eminently distinguished. . . . The establishment of a separate Professorship of natural and experimental philosophy and chemistry would also conduce essentially to a much more perfect course of instruction in the different subjects which it would embrace, some of which are at present entirely neglected.[57]

James Renwick, the young engineer chosen to teach the "neglected" sciences provided many of the early modern textbooks for American college science. The following year, 1821, Columbia became one of the earliest schools to require algebra as well as arithmetic for admission. But burdened by increasing library expenditure and Renwick's demands for

Robert Adrain Prof Nat Phil & Math

James Renwick

[55] Prof. [Samuel Latham] Mitchill, *The Present State of Learning in the College of New York* (New York, 1794), p.6. MS, anonymous student notes dated June 21, 1791, Columbiana Collection, Columbia University Libraries.

[56] *Report Relative to the Course of Instruction and Discipline in Columbia College, Presented to the Board of Trustees February 28, 1810* (New York, 1810).

[57] MS, Minutes of the Trustees of Columbia College, May 1, 1820.

better equipment, Columbia, heavily in debt, was forced to petition the state for relief in 1824.[58]

Columbia's request for public support brought the standard complaints about colleges into the open. By 1830, there was serious movement to start the rival institution, soon to become New York University. In response to the complaint heard on every campus about the "aristocratic" nature of contemporary education and out of anxiety over the frightening talk of a new college, on January 5, 1830, the trustees appointed a committee to draft "a plan for extending the scheme of instruction, and of establishing public lectureships." The finished plan called for a "Scientific and Literary Course, the whole or any part of which matriculated students may of their option attend."[59]

The plan did not work. Only Renwick's faithful execution of his public lectures enabled Columbia to get any "closer to the public" that year. When the alumni met in the spring, they voiced their strong support for a single liberal arts program.[60] By 1835, Renwick seemed to agree that growth had to be effected within the traditional college program. In his report, he admitted that partial students were "looked upon by the other students as of an inferior order" and noted that, although the "physical and mathematical departments [were] apt to [be] undervalue[d]," even regular liberal arts graduates of Columbia had become successful engineers and scientists.[61] Thus, when all Columbia's excursions from the regular college course (including the plan to instruct "young officers at the Navy Yard" in mathematics and science) were waning, there still remained the "full course" whose integrity, the trustees declared, "shall in all respects be preserved."[62]

Bowdoin College, too, was forced to ask in 1827, "Whether the courses of instruction ought not to be more of a practical

Bowdoin

---

[58] MS, A copy of the petition transmitted to the state legislature appears in the Minutes of the Trustees, Feb. 2, 1824. The trustees insisted that they could not raise tuition because a "great portion" of their students were poor New York youth, and that would place the "means of Education beyond [their] reach."

[59] MS, Minutes of the Trustees, Jan. 5 and 16, 1830.

[60] John T. Irving. *A discourse on the Advantages of Classical Learning, with a sketch of the Character of the late William Samuel Johnson delivered before the Association of the Alumni of Columbia College, at their Anniversary, 5th May, 1830* (New York, 1830).

[61] Renwick's report is a printed, two-page circular dated New York, December 16, 1835, in Columbia College Papers, Columbia University Library.

[62] MS, Minutes of the Trustees, Jan. 16 and Feb. 3, 1830.

and less of a scholastic character, and to this end whether the study of the Greek language in this college ought not to be optional with the student."[63] Never wishing to risk "anything ourselves by a hasty innovation," the College decided that it should always be constrained to the fixed "classical" curriculum.[64] The following year, however, Bowdoin did consider it safe to divide the single professorship of mathematics and natural philosophy held by Parker Cleaveland since 1808. Cleaveland relinquished the instruction of mathematics to William Smyth, who was soon instrumental in introducing modern French analytical thinking into the college.

Also concerned with changes in curriculum was Williams College, which, according to a modern historian, spent its largest energies prior to 1836 in just keeping open.[65] Following its meager beginnings in 1795, "for nearly twenty years no one class of studies, whether languages or mathematics or natural sciences or mental and moral sciences, seem to have taken precedence in interest over any other class of subjects taught." But finally, through the efforts of Amos Eaton, Chester Dewey, and Ebenezer Emmons, the students demonstrated a pronounced interest in science. A graduate of the class of 1799, Amos Eaton lectured at Williams in the spring of 1817; and the students, who delighted in his lectures, encouraged and even financed the publication of his *Manual of Botany*, dedicated to the "president and professors of Williams college . . . ."[66]

After completing its first subscription for the endowment of professorships by the end of 1826, the Williams trustees appointed a committee "to examine and report what changes are necessary in the present system of Collegiate Education." The committee recommended that political economy be immediately "added to the prescribed course of studies," that Hebrew be presented as an option, and that "attendance upon any of

[63] Report of the Visiting Committee quoted in *The History of Bowdoin College* by Louis C. Hatch (Portland, Maine, 1927), p.63.

[64] MS, Report of the Visiting Committee, Aug., 31, 1841, Bowdoin College Library.

[65] Frederick Rudolph, *Mark Hopkins and the Log: Williams College, 1836–1872* (New Haven, 1956), p.9.

[66] Arthur Latham Perry, *Williamstown and Williams College* (published by the author, 1899), pp.287–293, quotations from pp.287, 292. See also Frederick Rudolph, *Mark Hopkins and the Log: Williams College, 1836–1872* (New Haven, 1956), pp.138–140, for a description of early science at Williams.

the courses of Lectures be open to all persons qualified to profit thereby. . . ."[67]

Operating in this spirit of reform, Williams in 1828 selected, as its first lecturer on chemistry, Ebenezer Emmons, who became professor of natural history in 1833. The alumni met on September 5, 1832 for the express single purpose of "add[ing] to the Philosophical and Chemical Apparatus of the College" and were able to raise $4,000 to send an agent to Europe to make purchases.[68] By 1838, two years after Mark Hopkins became its president, Williams had established one of the earliest college observatories and had created separate professorships of mathematics, natural philosophy, chemistry, and natural history, comprising four out of seven positions on the professorial staff.[69]

Union, which like Williams opened in 1795, included in its original by-laws provision that "those students who do not wish to go through the regular course of studies required in this college may be permitted to study select parts of science [knowledge] with either of the classes." By 1802, however, only a few "partial" science students were stimulated by the scientific activity around Albany. It was not until after the almost immortal President Eliphalet Nott had assumed control of the college (1804–1866) that interest in science increased. Soon, the Erie Canal surveys begun in 1808, the appointment of Ferdinand Rudolf Hassler as professor of mathematics and natural philosophy in 1810, the opening of the Albany Academy in 1815, and the establishment of Rennselaer Institute in 1824 gave an impetus to "systematic and experimental science rather earlier at Union that in most liberal arts colleges."[70]

In 1828, when Nott added to his duties the presidency of

[67] MS, Minutes of the Trustees of Williams College, Dec. 5, 1826; May 9 and Sept. 5, 1827.

[68] Printed Circular of the Alumni Meeting dated Sept. 7, 1832. Miscellaneous Manuscript Collection, Williams College. MS, Minutes of the Trustees, Sept. 4, 1832, and Aug. 19, 1834.

[69] *Catalogue of the Corporation, Officers and Students of Williams College, 1838–9* (Troy, N.Y., 1838).

[70] Dixon Ryan Fox, *Union College: An Unfinished History* (Schenectady, 1945), *passim;* quotations from pp.14, 39. Warren Crosby Taylor, typed MS, "Beginnings of Engineering Education in U.S." (Feb. 4, 1955) at Union College Library. On Rennselaer, see Samuel Rezneck, *Education for a Technological Society: A Sesquicentennial History of Rennselaer Polytechnic Institute* (Troy, N.Y., 1968).

Rennselaer, he established at Union a firm foundation for the parallel course by creating a separate degree in science. After a year of study, a student could elect to pursue the "classical" course, in which more science and modern language was required at the expense of Latin and Greek. It should be pointed out, however, that even the "classical" students at Union could not avoid studying science. When they formed a committee in 1835 to request literary electives in place of optics and astronomy, the trustees denied that privilege on the grounds that the "standard of attainments must yet be elevated . . . in Optics and Astronomy."[71] With a student body then the third largest in the country, Union was successful in its plan of reform and by 1839 had one of the nation's largest teaching staffs and a student body surpassed in size only by Yale's.

Of all the northeastern colleges of the period, Middlebury was the most disadvantaged. Established in 1800, its first descriptive catalog of 1821 lists only four professors (including the president).[72] In 1828, the first year that the science department was fully represented, there was still only a president, a professor of languages, a professor of chemistry and natural philosophy, and a professor of mathematics and natural philosophy; and the men who filled these positions were ministers of no reputation. This unfortunate situation would not be altered at Middlebury until Alexander C. Twining and Charles B. Adams, two exceptional scientists, came to the college in 1839, some years after the other colleges were already more firmly established. Still, the school's pretensions remained small and as late as 1860 Middlebury, alone of all the colleges, did not yet have a separate professor of mathematics. Even a subscription started in 1833 for $30,000, some of which would be used to increase "the facilities of studying the natural sciences," was not filled by 1850.[73]

At midcentury, former President Joshua Bates (1818-1839) admitted that "Middlebury College labored long under the disadvantages of a course of study and instruction, too circum-

[71] MS, Minutes of the Trustees of Union College, July 22, 1835.

[72] In *Town Father: A Biography of Gamaliel Painter* (New York, 1952), p.239, W. Storrs Lee, historian of Middlebury, explains that the major portion of one of the founders' bequests was diverted from construction to the endowment of the professorship of mathematics and natural philosophy in 1819.

[73] MS, Records of the Corporation of Middlebury College, Aug. 22, 1833.

scribed; with a corps of officers too small, and means of improvement and instruments of illustration too limited."[74] Significantly, 397 of Middlebury's first 876 graduates (1800–1850) became ministers, a percentage far exceeding that for any other school.[75]

After the Supreme Court decision of 1819 silenced the political struggles which hindered Dartmouth's advancement, it too showed the progressive tendency of the age. James F. Dana and others connected with the Dartmouth medical school succeeded in interesting college students in chemistry and anatomy, until finally, in 1825, these lectures were required for the junior and senior classes.[76] However, a student writing to his former academy principal in 1823 could remark that the "college is very poor indeed."[77] The following summer, violent student rebellion erupted in which the president was hanged in effigy, a barn burned, and a professor stoned.[78]

When the new college laws appeared in 1828, they called for decreased domination by the trustees and for instruction in political economy and "Experimental Physiks," an area which had been sadly neglected.[79] Students, in fact, often entered Dartmouth with the intention of switching to Yale after two years for "the advantage it afforded in its *Scientific* Department."[80] Incredibly, Dartmouth had not subscribed to a scientific periodical in a half-century.[81] That summer, Frederick Hall, a Dartmouth alumnus and pioneer science educator, delivered the Phi Beta Kappa address at commencement,

---

[74] Untitled address by Joshua Bates in *Addresses and Proceedings of the Semi-Centennial Celebration of Middlebury College held at Middlebury Vermont, August 20, 21, and 22, 1850* (Middlebury, 1850), pp.17–40, p.29.

[75] *Ibid.*, p.32. Compare, for instance, the figure of 240 out of 1032 graduates of Bowdoin (1806-1850), the other isolated stronghold of Congregationalism, in Egbert C. Smyth, *Three Discourses Upon the Religious History of Bowdoin College during the Administration of Presidents M'Keen, Appleton, & Allen* (Brunswick, 1858), p.80. The records of Middlebury are quite incomplete and I am unfamiliar, for instance, with the nature of the conflicts which resulted in its huge faculty turnover. The only history of Middlebury is full of reminiscences and not generally helpful: W. Storrs Lee, *Father Went to College: The Story of Middlebury* (New York, 1936).

[76] MS, Minutes of the Trustees of Dartmouth College, Aug. 23, 1825.

[77] MS, D. Pillsbury to E. B. Moore, Dec. 27, 1823, Dartmouth College Archives.

[78] Leon Burr Richardson, *History of Dartmouth College* (Hanover, 1932) **1:** p.380, has explanation. Details of the action can be found in student's letter to his brother, MS, J. Willard to Augustus Willard, Aug. 18, 1824, Dartmouth College Archives.

[79] MS, "Laws," dated Jan. 21, 1828, Dartmouth College Library.

[80] MS, O.P. Hubbard to B. Silliman, April 23, 1836, Dartmouth College Library.

[81] Baxter Perry Smith, *The History of Dartmouth College* (Boston, 1878), p.278.

deploring the fact that "the sciences are too imperfectly taught;—too feebly patronized in our land." He recognized that it is a "fault which pervades our entire country;—which exists, in a greater, or less degree, at all our colleges . . . ," but hoped that Dartmouth would soon change.[82]

Shortly after, President Nathan Lord summarized the nature of the intellectual struggle in his inaugural address:

> The mass of the community becoming more enlightened, will call for proportionally higher qualifications in those, who are sent out to preside over the public interests, and their progressive influence will produce a yet more powerful reaction. But to meet these demands, amidst the conflicting sectional interests and fluctuations of public feeling, which are usually attendant upon a state of freedom; to discriminate rightly between the diverse systems of instruction and discipline, which are set forth with such frequency, and such earnestness of commendation; to keep so near the public sentiment, as not to lose the confidence of the community, and yet, not to follow it so implicitly, as to sacrifice the more desirable good of Self-Approbation [is the problem].[83]

In other words, Lord was not going to teach "mercantile, mechanical or agricultural operations," but he would not withhold support from science.[84] In 1836, bidding for Oliver Hubbard, Silliman's student and son-in-law, Lord offered "every possible facility in reference to chemistry and geology" and an initial appropriation of $700 for chemical equipment alone.[85] When Hubbard hesitated, Lord added, "I will take care that the laboratory shall be prepared for you as you request. Whatever purchases you may think necessary for the contemplated course of Lectures in April, you will please to make, & the bills shall be immediately paid by the College."[86] Hubbard was obviously impressed with the fact that the salary of his professorship was to be raised from $725 to the same

---

[82] Frederick Hall, *An Oration on the Importance of Cultivating the Sciences* (Baltimore, 1828), p.7.

[83] Nathan Lord, *An Address Delivered at Hanover, October 24, 1828, at the INAUGURATION of the Author as President of Dartmouth College* (Windsor, Vt., 1828).

[84] Richardson, *op. cit.* 1: p.389.

[85] MS, Nathan Lord to Oliver Hubbard, Feb. 27, 1836, Dartmouth College Library.

[86] *Ibid.*, March 29, 1836.

$1,000 the other professors received, that a professor had been dispatched to Europe to procure foreign journals, and that the average age of the entering student is now "18 to 20 years."[87] By 1838, Dartmouth had a separate professor of mathematics and two science professors—Stephen Chase, Oliver Hubbard, and Ira Young.

At least four schools—Brown, Princeton, Dickinson, and Pennsylvania—were in desparate circumstances by 1827-1828. Prior to the arrival of reform President Francis Wayland, the science curriculum at Brown had shown small change since the 1783 College Laws were enacted. In fact, fluxions, which were topics for commencement theses there in the late eighteenth century, "die away in the early nineteenth century."[88] By 1826, John D'Wolf, a professor who had studied under Robert Hare at the University of Pennsylvania Medical School, complained that other professors were "negligent, or uncompetent, or busied about other things to the neglect of [teaching]." He then continued to recount Brown's whole sad situation:

> Reform is indispensible . . . to the very existence of this Institution. First, Raise the requirements for admittance. Raise them *immediately,* & to an equality with those of our best Institutions—let it no longer be said (as it now is said with truth) that the rejected applicants of other colleges find ready admittance at our. . . . Second, Away with some of our schoolboy studies, & introduce others of a more manly character —geography:—this is not a college study—it belongs in the Academy—they should know it before they come to college. . . . Arithmetic too—examine them rigidly in this [for admission] and let it no longer be said that simple addition (though a very useful rule) is one of the classics at Brown. Third, Introduce the study of Political Economy . . . Fourth, Give them the option between the ancient & modern languages . . . let those who

[87] MS, O.P. Hubbard to B. Silliman, Feb. 27, 1836, Dartmouth College Library.

[88] Donald Fleming, *Science and Technology in Providence, 1760-1914* (Providence, 1952), p.27. It is interesting to note that although the curriculum remained static, student interest in modern subjects, including science, expressed itself in student societies. Professor Frederick Rudolph of Williams College has suggested that these student societies tend to indicate what will come about and then die away (personal interview, June 8, 1968). For support of this thesis in the case of Brown, see J. Walter Wilson, "The Philophysian (Philophysian) Society at Brown University (1818-1827)," *Books at Brown* **20**(1965): pp.41-55.

wish to be classical [gentlemen] have all facilities, [but] give the whole course a more useful & practical bearing.[89]

Wayland, the new president, had taught natural philosophy at Union College, where he had absorbed President Nott's personal love of science. An educator who was to call the Vermont report "by far the best thing on college education that I had seen," he immediately set out to correct Brown's deficiencies.[90] Algebra was added as an entrance requirement in 1828, and by 1829, Wayland said of Brown that "its requirements for admission are equal to those of the oldest colleges in N. England; and its course of study as elevated; and that its officers devote themselves exclusively to the business of teaching."[91] In 1850, when Wayland thought he had exhausted the possibilities of required science in the classical curriculum and wanted a free elective system and more practical instruction, he had four scientists on his seven-man faculty.

As late as 1821, Princeton's faculty consisted of the president, two professors, and three tutors.[92] Princeton, generally a school for wealthier students, was not getting enough of them. By 1827-1828, it "seemed that the end of the [College] was close at hand."[93] The trustees considered raising the tuition and even lowering the salaries of what professors they had. Finally, they decided that "if the college is to be supported at all, it must . . . render the course of the instruction, and the order of the institution such as to command a high degree of public confidence, & thus attract a larger number of pupils."[94] They did lower tuition and, after a few years of mediocre and short-lived appointments, assembled one of the most distin-

[89] MS, John D'Wolf to [President Asa Messer?], Aug. 12, 1826, Brown University Archives. D'Wolf gives evidence in this letter that he knew what was going on at Amherst.

[90] Julian Ira Lindsay, *Tradition Looks Forward, The University of Vermont: A History, 1791-1904* (Burlington, 1954), quoted on p.139.

[91] MS, Copy of a letter sent by Wayland to Harvard to request a Phi Beta Kappa chapter, 1829, in Faculty Minutes, Brown University Archives. A description of all the changes at Brown, including Caswell's elation over the "very extensive and superior philosophical apparatus with which the University has recently been furnished," may be found in *Annual Report of the Faculty of Brown University to the Corporation of that Institution* (Providence, 1829).

[92] *Catalogue of the Officers and Students of Nassau Hall, December 1821* [no publisher].

[93] Thomas Jefferson Wertenbaker, *Princeton, 1746-1896* (Princeton, 1946), p.179.

[94] MS Minutes of the Trustees, Sept. 27, 1827; April 8, 1828, Princeton University Library.

guished faculties in America. The faculty of seven, which included Joseph Henry and John Torrey, had four professors of mathematics and sciences and a language professor who also lectured in natural history. The 1832–1833 catalog asked candidates to note "that there has been a considerable reduction in the charges for board, tuition, &c., and that the course of study, and the provisions for conducting it, are more extensive than they have ever been."[95]

Dickinson, with a continuing scientific tradition invigorated by the addition of Joseph Priestley's friend Thomas Cooper as professor of chemistry (1811-1815), also had a troubled decade from 1821 to 1830. There was only one scientist on the small faculty of four—Henry Vethake, who was "Professor of Mathematics, and Natural Philosophy; [and] who also teaches Geography, Chemistry, and Political Economy."[96] The faculty, therefore, in 1828 petitioned the trustees to create a separate professorship of chemistry and natural history.[97] The trustees hesitated, and the entire faculty, no longer willing to accept the fact that it had no power on any matters concerning the college, resigned.[98] But the trustees did not learn a lesson. In 1830 they hired a new faculty which included Henry D. Rogers, later to become the most respected dynamical geologist in America at the University of Pennsylvania; but he was soon fired for "differ[ing] from the number of the Faculty about the concerns of the College, and its managements."[99] Again, the faculty dispersed. "When the college closed, in 1832, its funds were exhausted . . . [and] there seemed no hope of continuance."[100]

But hope for Dickinson came from a new source. The Philadelphia and Baltimore Conferences of the Methodist

---

[95] *Catalogue of the Officers and Students of the College of New Jersey; for 1832 & 1833* (Princeton, 1833), pp.14-15.

[96] *Catalogue of the Officers and Students of Dickinson College* (Carlisle, 1823), p.3.

[97] MS William Neill (Pres. of Faculty) to President of the Board of Trustees of Dickinson College, March 22, 1828, Dickinson College Library.

[98] Charles F. Himes, *A Sketch of Dickinson College* (Harrisburg, 1879), p.53. Vethake, who went to Princeton, printed his own explanation, *A Reply to A Narrative of the Proceedings of the Board of Trustees of Dickinson College, from 1821 to 1830* (dated December 15, 1830; copy in Dickinson College Library). The history of these strained faculty-trustee relations may be found in Saul Sack, *History of Higher Education in Pennsylvania* (Harrisburg, 1963) 1:pp.65-67.

[99] MS, Minutes of the Trustees of Dickinson College, July 10, 1831.

[100] William W. Edel, *John and Mary's College over Susquehanna* (New York, 1956), p.19.

Church, which were looking for a college to sponsor, replaced the old trustees within the framework of the original charter, supplied an endowment, and gathered a vigorous new faculty, headed by President John Price Durbin, who had been the first science professor at Wesleyan. The new plan called for six faculty members, including a professor of mathematics and a professor of science.[101] By 1836, mathematics was studied during each of the four college years and science during the last two—a plan "substantially the same with those of the oldest Colleges in our country."[102]

At the University of Pennsylvania, an older school in which science had always been held in some esteem, the college was not going well. Many of William Smith's original liberal "concepts may have influenced the Colonial curriculum, [but] they did not take root, and were subordinated to the interests of the dominant liberal arts program."[103] The University's famous medical school, however, kept alive the hope that the college would some day respond to scientific needs. In 1816, a plan to establish a botanic garden flowered into a proposal for a separate undergraduate department of "Natural Science," consisting of four professors. Like their counterparts elsewhere, the professors appointed were underpaid and underencouraged; and the plan officially ended when the faculty dissolved in 1827.

Reform of a more immediate nature had begun as early as 1813, when Pennsylvania's new provost toured the neighboring colleges, returning to recommend drastic changes at his own school. By 1825, the age of admission had been raised to fifteen, the standards of admission considerably increased, the academy separated from the college, and the college program extended to four years.[104] But the problem which still remained at Pennsylvania was that of trustee-faculty relations, which threatened to close this college just as they were

[101] MS, The new trustee committee's report on the establishment of professorships, Sept. 25, 1833, Dickinson College Library.

[102] *Catalogue of the Officers and Students of Dickinson College, Pa.*, (Baltimore, 1836), p.9.

[103] Saul Sack, *History of Higher Education in Pennsylvania,* (Harrisburg, 1963) **2:** pp.454.

[104] See Edward Potts Cheyney, *History of the University of Pennsylvania, 1740-1940* (Philadelphia, 1940), chap. 5, pp.176-216, entitled "Low Water 1791-1828" for an exposition of the problems which plagued all of the colleges during this "not [so] great academic age" (p.176).

destroying Dickinson. In the spring of 1828, Nicholas Biddle headed a committee of the trustees which finally recommended that the entire faculty be dissolved and that the college start afresh.[105]

The new faculty at Pennsylvania included Robert Adrain, the most gifted mathematician in America at the time, and Alexander Dallas Bache, the young scientific wizard from West Point. In his inaugural address, the new provost insisted that "Pennsylvania is an institution as broad in its principles, and as comprehensive in its course of instruction, as any college within our common country." While he was making these remarks, Pennsylvania's agent was en route to Paris to purchase $2,000 worth of physics and chemistry equipment. In July of 1829, President Andrew Jackson, Vice-President John C. Calhoun, and Chief Justice John Marshall attended the ceremonies for a building to be "adorned by increased apparatus for instruction in the sciences."[106]

Two years after the new program had gone into effect, the student body had increased from 21 to 126.[107] By 1836, when geologist Henry Darwin Rogers joined the faculty, three of Pennsylvania's six professors taught mathematics or science, mathematics being studied every year and science in the last three.[108]

Wesleyan University sprang up amidst all of the older colleges in 1831 and provides a convenient test of the new atmosphere their combined activities had created. Ironically, it was a Methodist minister and classicist, Wesleyan's first president, Willbur Fisk, who resisted all attempts to give the new college a theological flavor. He asked,

> Do the wants of the world need to be extensively consulted by none but ministers of the Gospel? It is a shame that this sentiment so generally prevails. The world wants farmers and mechanics [and] teachers and physicians for missionaries; it

---

[105] MS, Minutes of the Trustees of the University of Pennsylvania, July 2, 1828.

[106] William H DeLancey, *An Address delivered before the Trustees, Faculty, and Students, of the University of Pennsylvania on opening the Collegiate Session of 1830-31 in the New College Hall, on Saturday, September 18th, 1830* (Philadelphia, 1830), p.7. This report contains complete description of the changes that had taken place in the previous two years.

[107] *Ibid.*, p.49.

[108] *Catalogue* . . . .

wants statesmen and lawyers, yes, do not smile at the idea, it wants lawyers for missionaries.[109]

Fisk insisted that "modern literature, the natural and exact sciences, and the application of the sciences to the useful arts [are] first in importance in a useful education."[110] The school's first catalog outlined a "partial course" program, and by 1836, Wesleyan had spent $6,000 on scientific equipment.[111] By 1840, Wesleyan's faculty consisted of six professors and a president, with positions in natural philosophy and astronomy; natural science, and mathematics and civil engineering.[112] Wesleyan had proven that all colleges, old and new, "are generally awake to the importance of a thorough improvement, if not a radical change."[113]

As had all those colleges before it, Wesleyan had asked the question of how many new things might be taught in college. But, as one observer remarked, "The question, whether *a proper university* might extend its departments of instruction to every thing that needs to be taught among us, would be a very different one from the question, whether our colleges, arranged as they have been, and still are, can do it with the prospect of success."[114] Their conclusion to extend the courses in science guaranteed that for the next few decades, at least, the classical colleges would provide a significant home and nursery for national science.

---

[109] Wil[l]bur Fisk, *The Science of Education: An Inaugural Address delivered at the opening of the Wesleyan University in Middletown, Connecticut, September 31, 1831* (New York, 1832), p.7.

[110] *Ibid.*, p.13.

[111] *Wesleyan University: Catalogue* . . . (Middletown, 1831 and 1836–37). The *Catalogue* of 1836–37 (p.22) gives expenses of $6,000; Fisk's own records show "Apparatus $4,771.04, Laboratory, $397.57, and Cabinet of N. History $540.93" (MS, Willbur Fish [1836], Wesleyan University Library).

[112] *Wesleyan University: Catalogue* . . . (Middletown, 1840–41)

[113] Fisk, *op. cit.*, p.3.

[114] [Moses Stuart] "Study of the Classics," *Amer. Jour. Education* 3(1828): p.66.

III. Mathematics

Although mathematics was firmly established in the American college by the middle of the eighteenth century, little progress had been made in the means of its instruction. The only American mathematics textbooks were a few elementary arithmetics. Sometimes, arithmetics with some algebra could be obtained at reasonable prices from England or Scotland. But essentially, instruction depended upon the knowledge of individual teachers whose manuscript copybooks were used by students to produce their own copies. At worst, "mathematics" consisted of common arithmetic and a little surveying; at best, there were algebra and fluxions, or the elementary principles of the English system of the infinitesimal calculus.[1] After the chronometer replaced dead reckoning at sea in midcentury, navigation was included in the studies. But not until 1788 did an American textbook advance beyond arithmetic to the rudiments of algebra.

The causes of this low state of learning are varied. Not even the best European mathematicians of the eighteenth century had much effect upon their colleges and universities. Besides the simple physical problem of insufficient textbooks, there was little intellectual concern with the subject. Only almanac-makers and later navigators had any particular need for mathematics. "The farmer and artisan needed merely the ability to add and subtract numbers, rarely to multiply, and almost never to divide."[2] Before the great French works appeared in the nineteenth century, the only Americans who recognized a need for sophisticated mathematics were those

[1] Florian Cajori, *The Teaching and History of Mathematics in the United States* (Washington, 1890), p.25, reports that there was no algebra in schools before 1786; but Lao Genevra Simons, *Introduction of Algebra into American Schools in the Eighteenth Century* (Washington, 1924), p.3, *passim*, gives evidence that it was taught at Harvard in 1739 and elsewhere before 1786.

[2] David Eugene Smith and Jekuthiel Ginsburg, *A History of Mathematics in America before 1900* (Chicago, 1934), p.2.

very few who were capable of reading the *Principia*. For others, the study of Newton meant the *Opticks*, with its easier non-mathematical formulation. The great American urge to view the "Book of Nature" was accompanied by a belief that the observer needed nothing but his own eyes.[3]

Yet it is too simplified just to insist that Americans "stressed practical uses" in their mathematics.[4] Even though the arithmetics commonly reflected the new numerical demands of trade and commerce, the college did not have these ends in mind. To them, mathematics represented a form of mental exercise—not an exercise in the use of logical systems, but in the use of rules and formulae. Certainly, the strange methods of Euclid, the elaborate analysis of the conic sections, or the obscure language of the fluxional calculus could not have seemed practical to anyone. A writer explained in 1814 that a "higher object is proposed, in the case of those who are acquiring a liberal education [than] merely to obtain such a knowledge of the *practical* parts, as is required for transacting business . . . [but] as the learner is supposed to be familiar with none of the sciences except arithmetic . . . frequent references are made to mercantile concerns . . . *merely* [my italics] for the purpose of illustration."[5]

The most sophisticated example of an eighteenth-century student notebook is that of Samuel Miller, who copied it while a senior at the University of Pennsylvania, 1788-1789.[6] The first of Miller's two volumes contains algebra as far as the solution of simultaneous equations of two unknowns, trigonometric solutions of right triangles, and twenty-six pages devoted to the parabola. There are numerous problems ac-

[3] Brooke Hindle, *The Pursuit of Science in Revolutionary America, 1735-1789* (Chapel Hill, 1956), p.328, discusses this point.

[4] Phillip S. Jones, "The History of Mathematical Education," *Amer. Math. Monthly* 74, No.1, Part II (January, 1967): p.49.

[5] Jeremiah Day, *Introduction to Algebra, Being the First Part of a Course of Mathematics, Adopted to the Method of Instruction in the American Colleges* (New Haven, 1814), p.3.

[6] MSS, "Mathematica Compendia, or A Short System of Mathematical Literature, as it is taught by Robert Patterson, A.M. Professor of Mathematics in the University of Pennsylvania . . . 1789," Rare Book and Manuscript Room, University of Pennsylvania Library. Lao G. Simons, *op. cit.*, p.24, thinks that this work would have been published if there had been a more lively publishing industry at the time. Samuel Miller, the student, later wrote *A Brief Retrospect of the Eighteenth Century* (1803); see pp.15-16 above.

companied by tediously drawn out solutions. The second volume consists of "practical mathematics"—surveying, geometrical construction, and navigation, the geometry being presented largely without proofs or even posterior justifications. A single fluxional discussion arises in conjunction with a length of arc determination; and the practical problems of surveying and navigation require the use of logarithms. Although more reliance is placed upon the student's problem-solving abilities than in most of these works, the whole lacks rigor and continuity.

In 1788, Nicholas Pike produced the first American textbook to include algebra and other topics found in the manuscript notebooks.[7] Complete with a short section on geometrical constructions and trigonometry, the whole 512-page volume probably represents the total mathematical knowledge of the average college graduate in 1800. It was heartily endorsed by professors and presidents from Dartmouth, Brown, Yale, and Harvard as "well adapted even for University Instruction."[8] Thus, "we may fairly conclude that modern mathematics was substantially unknown in America in the eighteenth century."[9]

Pike's text was succeeded by a wave of American arithmetics, many of which also carried short sections on algebra and geometry.[10] The American works all followed English models and even used the English system of grouping numbers by six's or millions rather than by the French system of three's or thousands.[11] But most important was the fact that Americans were realizing that mathematics was relevant to nature, that

[7] Nicholas Pike, *A New and Complete System of Arithmetic* (Newburyport, 1788). Its second edition was edited in 1797 by Ebenezer Adams of Dartmouth. Chester Dewey of Union College edited the fourth edition in 1822 and the fifth and final edition in 1832.

[8] Pike, *ibid.*, unpaginated advertisements included in volume.

[9] Smith and Ginsburg, *op. cit.*, p.64.

[10] An incomplete list is found in Cajori, *op. cit.*, pp.48–49 (footnote). A discussion of all American arithmetics before 1860 may be found in James M. Greenwood and Artemas Martin, "Notes on the History of American Text-Books on Arithmetic," *United States Bureau of Education: Reports of Commissioner of Education for 1897–98* (Washington, 1900), chap. XVII, pp.789–868. Prior to 1820, Pike's textbook was gradually replaced in colleges by the work of Harvard's Samuel Webber, *Mathematics . . .* (Boston, 1801) and by Charles Hutton [English], *A Complete System of Practical Arithmetic* (New York, 1810).

[11] Cajori, *op. cit.*, p.51.

they could "no longer elicit physical truth by a tedious cross examination of [their] treacherous senses."[12]

Finally, Jeremiah Day of Yale published the first volume in America devoted exclusively to algebra, effecting an immediate change in the pattern of collegiate instruction.[13] Day consulted English works patterned after the great mathematical revisionist Leonard Euler and also Lacroix's *Algebra*.[14] His was the first book to be written with the student in mind, and with the admission that "in the colleges in this country there is generally put into the hands of a class, a book from which they are expected of themselves to acquire the principles."[15] For Day, mathematics was not the memorization of rules. A textbook "should not be rendered obscure and forbidding by [its] conciseness." Mathematics could not hope to "awaken interest, by rhetorical elegance, by exciting the passions, or by presenting images to the imagination . . . . It is the *logic* of the mathematics which constitutes their principle value, as a part of a course of collegiate instruction."[16] Although Day's 296-page algebra lacks a sufficient number of problems, its clearness and logic cannot be denied. The modern reader would have great difficulty in trying to understand the object of a discussion in an eighteenth-century arithmetic, but he would have no problem following Day or subsequent authors.[17]

Day's algebra was immediately accepted by all the colleges. In some, this and his subsequent volumes on other mathematical subjects were used for two decades or even longer. Harvard, soon followed by the others, would be able to add algebra to its requirements for admission in 1820. American colleges had been introduced to a system in which memory was eliminated, conclusions justified, and parts systematically

---

[12] James Dean, *An Oration on Curiosity pronounced in the University of Vermont, 24th April 1810, on Induction into Office* (Burlington, Vt., 1810), p.19.

[13] Jeremiah Day, *Introduction to Algebra* (New Haven, 1814).

[14] *Ibid.*, revealed in the footnote to p.233.

[15] *Ibid.*, p.3.

[16] *Ibid.*, p.4.

[17] In one of the most popular early nineteenth-century arithmetics, Thomas Dilworth's *The Schoolmaster's Assistant: Being a Complete System of Practical Arithmetic* (Philadelphia, 1802), p.25, we read under "subtraction": "Q. What number must you borrow when the lower Number is the greater. A. The same which you stop at in addition." See Cajori, *op. cit.*, pp.48–52, *passim,* for other examples of the obscure rules of arithmetic.

related. A belated revolution had finally begun in America; but it would not be completed until the entire French mathematics, known only to a few Americans through journals or the French, was imported into the college. This was a task which would occupy the period 1818–1836.

It was through both American and English translations that French mathematics was made available to American college students. Although the Cambridge Circle had brought continental mathematics to England as early as 1813, America, which trailed about a decade behind in most topics, at first preferred its own French translations. Later, the modern works of Playfair, Bonnycastle, and Young found ready acceptance alongside these American translations.[18]

The first American translator was John Farrar of Harvard, whose translation of Lacroix's *Arithmetic,* obviously intended for college use, appeared in 1818. He translated part of Legendre's *Elements de Geometrie* (1794) in 1819; and in 1824 he produced his own calculus, the first in America to display the French system. Within a decade, he had translated about a dozen works on mathematics and science, a selection of which came to be known as the Cambridge (Mass.) course of mathematics. Following Farrar, other Americans, usually college professors, made their own translations and revisions of the French works. Soon, every publisher in the now booming American industry was trying to get a piece of the new market created by the French. The largest share of that market was finally, in 1834, captured for a generation by Charles Davies (1798–1876).

Davies had graduated in 1815 from West Point, where the instructors, trained at the Ecole Polytechnique, made this the first institution where the French influence was felt. Davies himself taught at West Point (1816–1837), at various other colleges, and finally at Columbia (1857–1865). His *Elements of Geometry and Trigonometry* (based on Legendre, 1834) was in its fifteenth edition by 1860 and was not obsolete until the

[18] It would become too unwieldly to enter upon a discussion of each French mathematician's improvements in theory or pedagogy, each American translation, or the date when each college adopted a particular text. Besides general histories of mathematics, the reader is referred to *Cajori, op. cit.,* chap. III, pp.98–292, and especially to Lao G. Simons, "The Influence of French Mathematicians at the end of the Eighteenth Century upon the teaching of Mathematics in American Colleges," *Isis* 15 (1931): pp. 104–123.

end of the century; while *Elements of Algebra* (from Bourbon, 1835) went through a new edition almost every year for two decades and was last edited in 1875. Davies also produced the first American analytical geometry, *Elements of Analytical Geometry* (1836), although this new French phenomenon was first introduced here through two translations in 1824 and 1833. While none of the colleges studied here used all of Davies's many books, most colleges used at least one of them. But whatever version of the new mathematics they selected for use, every college by 1836 had modern textbooks of algebra, geometry, trigonometry, analytical geometry, and calculus. By 1840 there were scores of new books; and the format of these basic texts would serve every college student for at least the next fifty years. Albert Dod at Princeton, who himself edited the American edition of J. R. Young's textbook of calculus, "regard[ed] the multiplication of textbooks . . . as affording cheering evidence that juster ideas are beginning to prevail in our country respecting the proper scope of mathematical education."[19]

Adoption of the new textbooks was not a difficult matter, since the value of mathematics in general had been long conceded. At most colleges, the change to modern texts occurred as one of the reforms effected after the critical appraisals of the late 1820's. Yet as early as 1820, when the process had only barely begun at Harvard and when there was still no textbook for either analytical geometry or calculus, Columbia called upon both Robert Adrain and newly appointed professor James Renwick to explain what was happening in mathematics and what Columbia ought to do about it.

Recently charged with mathematical instruction alone, Adrain, then the most skilled mathematician in America, suggested that algebra be placed immediately after arithmetic as "the surest and best method of rendering the Collegiate system of Studies commensurate with the present improved state of mathematical science as taught in the writing of the best mathematicians in Europe." He lamented "the very abstract manner in which the science of geometry is presented in the Elements of Euclid"; and although he thought the

[19] Albert Dod, "Analytical Geometry," *Essays, Theological and Miscellaneous . . .* (New York, 1847), p.466 [essay written in 1841].

newer geometries of Simpson and Playfair acceptable, he recommended the "Geometry of Legendre" or the "course of LaCroix." He then insisted that to read modern post-Newtonian "mechanical philosophy, it is indispensable to pursue the French system of variable quantities as taught by Bossuet, Garnier, LaCroix and others and particularly La-Grange." He concluded by "remarking that although it may be proper to adopt a selection of the best books to be studied in the mathematical department; yet nothing could be more injurious to the interests of the institution than to limit the course of instruction to the mere contents of any such selection."[20]

In another letter, Renwick outlined the recent history of mathematics at Columbia, showing it to be strikingly similar to that at the other colleges:

> The course of Mathematics laid down in the Statutes as they now stand, was founded upon the practice of the late Dr. Kemp [appointed 1786, died 1812] and commenced at a very early period of his connection with the college. I am well aware that he wished it were in his power to have made changes in it . . . but loaded as he was with the duties of two professorships . . . he made no change in his general plan, & contented himself with a few explanations of modern methods as far as circumstances brought them within his views. I am aware that Dr. Adrain [since 1813] has gone still further than he did, in his exposition of the methods & inventions of the Mathematicians of Continental Europe, but in so doing he certainly departs still more from the letter of the Statutes.

Renwick then suggested "the propriety of adopting by Statute that analytical method [Continental], through which the power of perusing the work of Laplace is to be attained, in preference to the Geometric." He noted that the "chief difficulty that will arise in making the change is the want of proper textbooks," but pointed out that translations had begun at Harvard, suggesting that Columbia continue the work and "be the first college in which the English language is used to adopt the unbroken series."[21]

---

[20] MS, "Report of Prof. Robert Adrain to Pres. Harris about studies, Nov. [ ], 1820," Columbia College Papers, Columbia University Library.

[21] MS, "Prof. Renwick's report to Pres. about studies, Nov. 13, 1820," Columbia College Papers, Columbia University Library.

In 1821 Columbia, joined Harvard in requiring elementary algebra for admission and in then proceeding with one of the modern French tracts—a pattern which was repeated elsewhere. At Dartmouth, the freshmen petitioned the faculty to abandon the old arithmetic that had been used and to adopt the new course, "the study of which has been attended both with more satisfaction and improvement . . . [at] another respectable and long established college."[22] A student at Bowdoin asserted "that the common systems of arithmetic cannot make the thorough mathematician. It is impossible that the reasoning faculties can be enlarged and envigorated, by being confined within the limits of a few formula."[23] By 1836, all colleges had adopted courses derived from the French. Most now required some algebra courses derived from the French. Most now required some algebra for admission and calculus for graduation. By 1840, all the schools had separate professors of mathematics.[24]

The dramatic change which had occurred in mathematics education in less than two decades can, of course, barely be grasped from the catalogs of these colleges. It was a change in philosophy and pedagogy as much as in subject matter. The whole French system was constituted to "better serve to fix the propositions in the minds of their students,"[25] providing "principles with only such examples as are in themselves instructive or illustrative of a general truth" instead of the mere "examples" of an earlier day.[26] In comparing the new analytical method to that used in the tedious propositions of the Fifth Book of Euclid, one professor remarked that "the student [in two or three lessons] acquires a more valuable knowledge of proportions than he could do by 20 recitations in the method of the Greek geometers."[27] It was "universally

---

[22] MS, "The Comitte [*sic*] [of freshmen] to the Immediate Government of Dart. College," Nov. 1, 1824, Dartmouth College Archives.

[23] Anonymous student, "Mathematics," *Escritoir* 1, No.10 (Brunswick, 1827): p.146.

[24] The only actual exception is Middlebury, which, at times, did have only one professor of mathematics and natural philosophy. Vermont, Amherst, and Brown retained the title, professor of mathematics and natural philosophy, even after their faculties included three or four scientists; and actually, they did so quite appropriately, since rarely did the mathematician not also teach astronomy or mechanics.

[25] Lao G. Simons, "The Influence of French Mathematicians . . . ," p.109.

[26] MS, John Farrar to President [Kirkland?] Jan. 22, 1817, Harvard College Archives.

[27] MS, E. Snell, "Report . . . ," June 26, 1829, Amherst College Library.

conceded that without the aid of modern analyses the science of geometry cannot be established upon a rational basis. And without the help of geometry, thus established and ordered, all the real sciences, excepting only those included in the department of natural history, must be deprived of their development and perfection."[28]

The change in course content was accompanied, predictably, by gradual changes in methods of instruction. Describing his course of mathematics at Bowdoin College, an 1820 graduate remarked,

> Of the methods of instruction . . . the only exception I remember to simple recitation was a single slight piece of surveying. We were required to study the prescribed lesson in the book, then to repeat it, not of course word for word, but distinctly, to the professor or tutor at the recitation; that was all.[29]

Naturally, students often kept notebooks in which they solved problems and which they showed to the teacher outside of recitation. In 1824, Bowdoin's William Smyth became the first college professor to use a blackboard,[30] introducing with that simple pedagogical aid a minor revolution. Before the blackboard, "each student had his slate, and when he finished his work he took the vacant chair next the teacher's and underwent examination of the process or principle involved."[31] It must have been more difficult for students to perform operations before the whole class, for in 1830 the sophomore class at Yale, finding its new mathematical chores too difficult, requested "permission to explain Conic Sections from the book, and not to demonstrate them from the figures."[32]

Black board

---

[28] Albert Dod, *op. cit.*, p.465. To get some idea of the difficulty of handling higher mathematics before the introduction of French analysis, see the first pages of the following two textbooks: Samuel Vince, *The Principles of Fluxions: Designed for the Use of Students in the University* [first American edition] (Philadelphia, 1812)—the first book on calculus available in America; and J. R. Young, *The Elements of the Differential Calculus* (Philadelphia, 1833)—one of the newer calculuses based upon the French system that appeared in the 1830's and that was used at Columbia, Union, Pennsylvania, and Princeton, among others.

[29] Quoted in Cajori, *op. cit.*, p.77.

[30] Alpheus Packard, *Address on the Life and Character of William Smyth, D.D.* (Brunswick, 1868), p.14. Cajori, *op. cit.*, p.170.

[31] Cajori, *op. cit.*, p.171.

[32] *A Circular Explanatory of the Recent Proceedings of the Sophomore Class, in Yale College* (printed New Haven, August, 1830), p.3. Copy in Yale Memorabilia Room, Yale University Library. I assume that this request was denied.

Eventually, students "had more frequent exercises than heretofore in the solution of original problems connected with the subjects pursued in the text, and with excellent effect."[33] The changed character of the recitation for student and professor alike may be seen in a Brown professor's description of the technique in 1844:

> I adopted . . . the procedure of giving out at each recitation 10 or 12 written questions to be solved by one portion of the class on paper while the others were occupied in viva voce recitation and at the Black Board. The preparation of these questions and the examination of the written solutions subsequently occupied at least 1½ hours per day over and above the ordinary preparation for class exercises.[34]

A student at Amherst in 1847 said the professor "is not content with hearing simply what is contained in the book, but goes into the principles, traces out a subject in all its connections, illustrates it in practice, etc."[35] "Written exercises were frequently required . . . and the study of the [mathematical] subjects beyond the limits of text books was encouraged by all the means in [the college's] power."[36] Every student now attended mathematics recitations an hour a day for either three or four years of college. Calculus, completely unknown to the student in 1820, was by 1840 received by the faculty with the same enthusiasm that had been reserved for theology in a former day.[37] At the University of Vermont when the calculus appeared in the first term of the sophomore year, "all sophomores who received the mark *seven* or under in their

[33] MS, Alexis Caswell to the Faculty of Brown University, July 25, 1846, Brown University Archives.

[34] MS, Alexis Caswell to [?], July 27th, 1844, Brown University Archives.

[35] William Gardiner Hammond, *Remembrance of Amherst: An Undergraduate's Diary, 1846–1848*, ed. George F. Whicher (New York, 1946), p.87. Student diaries provide valuable material for social history. Hammond's diary, the only published diary for this period, is a condensation of the original, a copy of which is now housed in the Amherst College Library.

[36] MS, Minutes of the Trustees of Columbia University, Jan. 4, 1837.

[37] Strictly speaking, this statement is not true. I have mentioned the scattered appearances of fluxions even in the eighteenth century; they were, however, an optional study in the early nineteenth century, if they appeared at all, and, I would imagine, a not too popular option. At Harvard, for example, where some calculus [fluxions] had appeared as an option, it was never "made a part of the stated course of studies" until 1824: MS, John Farrar to Aaron Hill, May 26, 1825, Harvard College Archives.

examinations in calculus, [were] required to attend an [extra] recitation" the following term.[38]

The average student, who in the 20's had welcomed a change from the boring mathematical dialogues of the past, soon began to "inquire, with evident distrust, what good can result from abstract mathematical processes."[39] A group at Dartmouth, after finishing the "difficult study [of] Davie's [*sic*] Mathematics" appointed a student "to deliver an Eulogy on said Davie [*sic*]."[40] A class at Amherst burnt "Playfair's *Euclid* on the green with groans and other appropriate ceremonies."[41]

A new and constantly plaguing problem had arisen with the vigorous mathematical training now required of everyone—namely, how to make all studies palatable to all students. The early answer to this question was merely a reiteration of the mental faculty ideal. Mathematical studies were considered especially good for developing the "judging and reasoning faculty" and every student was expected to go as far as calculus for a liberal education.[42] The professors of mathematics expounded the principle that the "discipline of the mind [is] at least, *as* important an object in instruction as [in] the communication of knowledge."[43] But many students did not like calculus, or analytical geometry, or the other subjects they were required to study. Vindication built upon the necessity of mathematics for a proper appreciation of physical science gained none of their sympathy. The archival remains of the 1840's are filled with students' requests to be exempt from calculus, to perhaps substitute some favorite

[38] MS, Faculty Meeting Records, July 29, 1843, University of Vermont Archives.

[39] Solomon Stoddard, in *Addresses delivered at the Inauguration of the Professors of Middlebury College, March 18, 1839* (Middlebury, 1839), p.14.

[40] MS, G. B. Barrow to J. S. Barrow [his father], June 13, 1839, Dartmouth College Library.

[41] Hammond, *Remembrance of Amherst . . .*, p.196.

[42] Alexis Caswell, "Importance of the Mathematical Studies, Considered As A Branch of A Liberal Education," *The Biblical Repository and Quarterly Observer*, No.XXI (January, 1836), p.7.

[43] MS, Ebenezer Snell, "Series of questions relating to amount of Studies, &c in College. Reported June 26th, 1829," Amherst College Library. Claude Moore Fuess, *Amherst: The Story of a New England College* (Boston, 1935), p.113 (footnote), related the story of Henry Ward Beecher who asked Snell why he had to study mathematics if he were "preparing to be a clergyman." When Snell replied, "To discipline your mind," Beecher responded, "If that's all I shan't go to class any more. My mind gets enough discipline inventing excuses to you for not being there."

study in history or modern language, or to review the earlier mathematics as an alternative. At least more algebra and geometry would be better preparation for teachers of mathematics "as [it] is required in many of our common schools." But who planned to use analytical mechanics?[44]

Naturally, most of these requests were initially denied. If the president consulted with the mathematics professor, he would inevitably be told that

> The [Mathematics] now embraced in the course I deem indispensable to a Liberal Education. I recommend therefore no omission. Nor in my opinion, can the course be materially abridged without injury to the student and detracting from the reputation of the University.[45]

But the problem of the difficulty of mathematics for so many students could not be wished away. Soon after calculus was required at Harvard, it was observed that "about one third of the class were making little or no progress."[46] By the 1840's Charles H. Davies could insist that inadequate textbooks and methods of instruction were no longer the problem. He attributed the two-thirds failure rate at West Point to the "generally unattractive" nature of mathematics for most students, suggesting that, if Harvard or the other schools ever upheld the same standards, they would get the same results.[47]

Thus, in mathematics, as in all subjects, the increased sophistication of knowledge would now require some adjustment in the reigning principle that all students should take exactly the same courses. In the late 40's and 50's, students were gradually granted options, which allowed some to escape the mathematical pressures that had built up beyond elemen-

---

[44] MS, 20 Members of the sophomore class to the Faculty of Dartmouth College, March 22, 1849, Dartmouth College Archives.

[45] MS, Alexis Caswell to Rev. Dr. [Francis] Wayland, July [ ], 1841 Brown University Library.

[46] MS, Professor [James] Hayward, "Annual Report," [Oct.] 1826, Harvard College Archives.

[47] MS, Charles Henry Davies to Edward Everett, Feb. 17, 1848 (in Overseers' Reports, Vol.8), Harvard College Archives. Davies's figures and explanation are substantially confirmed in *Report of the Commission appointed under the eighth section of the act of Congress of June 21, 1860, to examine into the organization, system of discipline, and course of Instruction of the United States Military Academy at West Point* (Senate Miscellaneous Document, No.3; Congressional Serial Number 1089), p.198 and *passim*.

tary calculus. But before "elective" thinking really became a working idea after the Civil War, there was a period in which mathematics, at a high level, was required of everyone. In present-day terms, the nineteenth-century curriculum had advanced from the eighth-grade to the second year of college level in a single generation, from 1820 to 1850. During that time justification for the study of mathematics was increasingly announced in terms of its "relation to an acquaintance with Natural Philosophy," rather than primarily its "Immediate discipline of the reasoning powers."[48] To teachers, at least, it was clear that the applicability of the subject lay not in navigation, commerce, or even mere reasoning, but in science.[49]

---

[48] A. Caswell, "Importance of the Mathematical Studies . . . ," p.4.

[49] This point is made throughout Charles Davies's manual for instructors, *The Logic and Utility of Mathematics with the best methods of Instruction Explained and Illustrated* (New York, 1850), p.322, *passim*.

IV. Physics

It is impossible to discuss physics in the nineteenth century without first discussing the contemporary usage of scientific vocabulary. In colonial times and into the nineteenth century the term natural philosophy, as used in the colleges, included physics, chemistry, and astronomy.[1] The word science, used to denote all knowledge, was similar in meaning to the German *Wissenschaft*.[2] Only after 1830 was the word science first applied in its present sense. By that time it was common to use the longer term physical science to denote natural philosophy, chemistry, and natural history. Physical or natural science, then, was almost as inclusive a term as natural philosophy had been. The usage is further complicated by the new word physics, which was equivalent neither to the French "physique" or the modern "physics," being a term used almost as freely as physical science. Although the new discipline did not usually include chemistry, it was common in 1850 to find lectures in zoology and physiology listed under lectures in "physics." According to the most common contemporary distinctions,

> *Natural Philosophy* may be considered, as the science which examines the general and permanent properties of bodies; the laws which govern them, and the reciprocal action which these bodies are capable of exerting upon each other, at greater or less distances, without changing their nature.

---

[1] Failing to appreciate the fact that natural philosophy was not philosophy as later understood, modern writers have obfuscated the transformations of science in the college. For instance, Howard Mumford Jones says that the "natural philosophy" course of the period, "unlike most modern science courses, was really philosophical": *Ideas in America* (New York, 1965), p.153. And in their historical introduction to *Origins of American Scientists* 1 *(New York, 1952), R. H. Knapp and H. B. Goodrich insist, "One might search in vain within the colleges and universities of that time [1850] for scientists in the modern sense. With few exceptions, science was taught, if at all, only as natural philosophy" (p.2).*

[2] John T. Merz, *A History of European Thought in the Nineteenth Century* (New York, 1965) 1: pp.89–93, has elaborate descriptions of the European usages of the words for science.

> *Chemistry,* which treats of the changes, which the ultimate atoms of dissimilar bodies exert upon each other, at insensible distances, which always result in the production of bodies possessing new properties.
> *Natural History,* which describes the obvious external character of bodies, and their peculiar properties, and classifies them according to certain resemblances and differences. It has many subdivisions, as Geology, Mineralogy, Ornithology, Conchology, &c.[3]

At midcentury, the divisions of natural philosophy were mechanics, hydrostatics, hydrodynamics, hydraulics, pneumatics, acoustics, optics, astronomy, electricity, galvanism, magnetism, and chromatics—terms which did not exactly correspond to present usage.[4] Textbooks in mechanics, for instance, almost always included hydrostatics, hydraulics, hydrodynamics, pneumatics, and even acoustics. After 1830, when a single professor and textbook no longer embraced all of science, as many as four separate textbooks were often required for the study of what was known as natural philosophy. There was always a separate textbook and course in astronomy, and at least two textbooks for the remaining physics curriculum—one in mechanics and one on optics. Electricity and magnetism, knowledge of which was growing faster than that of the other subjects, were most often taught by lecture.

Despite the colonial enthusiasm which first brought science into the college curriculum, little was done in the eighteenth century to provide adequate instruction. At the beginning of the nineteenth century, "in mathematics and physics there was extremely little learning in the country. There were probably not half a dozen persons in New England who knew the rudiments of the differential calculus."[5] For the men who delivered the natural philosophy lectures in the colleges, their work must have been a labor of love, for they were usually not members of the regular faculty and often not paid. One such "professor" appointed at Brown in 1786 received no salary at

---

[3] MS, Rob't L. Dashiell, "Professor Allen's Lectures on Natural Science, 1845-6," Dickinson College Library.

[4] *Ibid.* I will use the term physics to apply to all of these except astronomy, treated separately in the following chapter.

[5] John G. Palfrey, *Notice of Professor Farrar* (Boston, 1853), p.9.

all until 1794. He complained that his was "an office which was but a name, without an apparatus." The Corporation then awarded him $100 a year salary, some of which he used to buy equipment which, at his retirement in 1803, the Corporation confiscated without compensation.[6]

Lectures in natural philosophy, when delivered, did not reflect the contemporary state of science. Early in the century when Jeremiah Day of Yale had to deliver the physics lectures he admitted that he did not have enough "time to canvass the subjects, and render them familiar [to himself]"; consequently, his lectures were a "contemporaneous *hodgepodge*."[7] It must be admitted, however, that improvement was effected in the first two decades of the nineteenth century. Although the professor of science continued to receive a smaller salary than his classical colleagues, colleges no longer allowed these professorships to remain vacant for years. The professors appointed in the 1800–1825 period were usually clergymen whose only scientific background was some enthusiasm left over from their own college days. A few, like John Farrar at Harvard, soon recognized the antiquarian nature of American physics and made an effort to import recent European science. But nevertheless, from 1790 to 1830, almost without exception, college students in America had their first and last look at physics through recitations from Enfield's *Institutes of Natural Philosophy*.[8]

The English Enfield had apparently had to write a book on natural philosophy in order to complete his repertoire of publications on history, biography, elocution, metaphysics, sermons, and hymnals. It was common practice in Europe, and later in America, to write science schoolbooks with very little knowledge of science.[9] This, of course, meant that writers composed with nothing more than scientific treatises

---

[6] MS, Peres Fobes to the Corporation of R. I. College, Sept. 1, 1803, Brown University Archives.

[7] MS, Jeremiah Day to Benjamin Silliman, March 12, 1804, Stokes Autograph Collection, Yale University Library.

[8] William Enfield, *Institutes of Natural Philosophy, Theoretical and Practical* (2nd American ed., Boston, 1811; first London, 1783; second London, 1799; first American, 1802; third American, 1820). I have consulted only the second American edition, 1811, to which I refer.

[9] This procedure was probably somewhat curbed in America by the fact that it was always cheaper to reprint an English work than to pay the price of the plagiarists' new editions.

and blank paper. Enfield, however, according to his own words, chose the works of "Newton, Keil, Whiston, Gravesande, Cotes, Smith, Helsham, Rowning, and lastly Rutherforth," most of which were written in the first half of the eighteenth century.[10]

The book begins with a short chapter on matter, echoing the enthusiasm of the old authors who "were building systems of principles for a world of material substances which existed only in their own imaginations."[11] The format of propositions and scholia is copied exactly from Newton's *Principia;* and its propositions involve the tedious geometrical reasoning of an earlier century.[12] The section on optics is transcribed principally from Newton's *Opticks;* it asserts that "rays of light consist of particles of matter,"[13] and does not even mention alternative interpretations. The short sections on electricity and magnetism, which give the appearance of an afterthought, are not connected and do not contain a mathematical formulation of any kind.[14]

A reviewer in the *American Journal of Sciences,* enraged when the third American edition of the "ill-digested compilation" appeared in 1820, wrote:

> For more than twenty years past, the compilation of Enfield has been extensively employed in this country as a textbook in Natural Philosophy; and at present, we believe, is used for this purpose by nearly all the higher seminaries of learning in New England. . . . The Institutes of Natural Philosophy will be found in examination to be little more than an abridgement, without much alteration either in language or form from different works in the same subject which appeared in the former part of the last century . . . The compiler of this work, while sitting with his pen in one hand and his author in the other, appears never to have indulged for a moment the illiberal suspicion that his author might be in the wrong. The consequence is, that whenever his original erred, the error is

[10] Enfield, *Institutes,* p.vii.

[11] "Remarks on Dr. Enfield's *Institutes of Natural Philosophy*—Third American edition—1820," *Jour. Science,* 3 (1821): p.128.

[12] E.g., "The velocities with which bodies move, are directly as the spaces they describe, and inversely as the time in which they describe these spaces." *Institutes,* p.13.

[13] *Ibid.,* p.165.

[14] See the following chapter, pp.80-82, for criticism of astronomy in Enfield.

> faithfully transcribed into his abridgement . . . and hence presents scarcely any idea of the progress made in the different branches of philosophy since the period of Newton . . . . The names of Bossut, Coulomb, Hutton, Dalton, Biot, Young and other great experimentalists who have almost created a new era in philosophy by their success in gleaning up the scattered laws behind them . . . scarcely once occur in a work which embraces nearly the whole extent of the philosophical studies in our system of liberal education.[15]

The copy of Enfield owned by Robert Adrain while he was at Rutgers is full of marginal corrections and sarcastic remarks. An exceptional mathematician, Adrain wrote next to several passages that they are "exceedingly tedious."[16] But by 1821 it appears that for even the less skilled readers

> The want of an elementary system . . . exhibiting a more faithful outline of the existing state of physical science, is beginning to be generally felt; . . .[17]

As inadequate as Enfield's text was, it was not, however, intentionally simplified or watered-down as were the popular volumes which existed alongside college textbooks after 1830. Rather, it represented the kind of academic science one would have expected to have heard a century earlier or even as late as 1780, but certainly not in the nineteenth century. On the eve of discarding Enfield from the Yale curriculum, Denison Olmsted could note "that so many fields of research have recently been opened, and so many new instruments of inquiry been put into our hands."[18]

There is, of course, some variation in the dates on which individual colleges adopted new material. However, we can differentiate three new stages of development in physics textbooks prior to 1860. The first is the period from 1825 to

3 stages in dev of physics textbooks

---

[15] *Amer. Jour. Science* 3 (1821): pp.125–129. The author is careful to use the expression "nearly the whole extent," probably because by this time there were separate chemistry texts in use. Samuel Webber, president of Harvard, who edited the 2nd American edition (1811), in fact, omitted the section on chemistry because it would not be used in American colleges.

[16] Copy dated August, 1812, now in the Rare Book Room, Library of the University of Pennsylvania.

[17] *Amer. Jour. Science* 3 (1821): p.125.

[18] Denison Olmsted, *An Oration on the Progressive State of the Present Age* (New Haven, 1827), p.17.

1833 during which Enfield was replaced at all colleges by three newer American textbooks and during which separate texts on astronomy and on mechanics also appeared. The second period, 1833–1850, is marked by the use of a separate textbook on optics and the extension of teaching by "experimental lecture," especially lectures on electricity, magnetism, meteorology, and the steam engine. The third period, dating after 1850, is that in which the textbooks finally presented mechanics and contemporary electricity and magnetism in sufficient mathematical detail.[19]

James Renwick at Columbia, John Farrar at Harvard, and Denison Olmsted at Yale, working generally from French treatises, provided the American colleges with the first three textbooks at all relevant to the times in which they were written. They replaced the single volume for all physical science with a single volume on mechanics alone.

Renwick, who had studied in Europe after his graduation from Columbia College in 1807, returned with a knowledge of French and German and a determination to teach modern science. He began lecturing at Columbia in 1814 and was appointed professor of natural philosophy in 1820, one of the first men to hold that office without the encumbrance of many other duties. His lectures on mechanics, astronomy, and the remainder of physics were published in two volumes between 1823 and 1826.[20] Samuel Miller, New York doctor and author who had experienced the old eighteenth-century science curriculum at the University of Pennsylvania, read the first volume and wrote to Renwick:

> I have perused the work with peculiar satisfaction; and think it much the most profound, clear, rich, and instructive compend of the subject of which it treats, that I have ever seen.[21]

---

[19] There is not sufficient space or need here to relate in detail the exact contents of study in each division of physics. A recent doctoral dissertation has carefully analyzed the course content at Princeton during the tenure of Joseph Henry and I have found the course to be substantially the same at other institutions. See Charles I. Weiner, *Joseph Henry's Lectures on Natural Philosophy: Teaching and Research in Physics, 1832-1849* (Ph.D. thesis, Case Institute of Technology, 1965), pp.68-237. Weiner also provides examples of examination questions in physics, pp.75-77.

[20] James Renwick, *Outlines of Natural Philosophy: being the Heads of a Course of Lectures; delivered in Columbia College, New York,* (2 v., New York, 1823, 1826).

[21] MS, Samuel Miller to James Renwick, Feb. 28, 1823, Renwick Family Papers, Columbia University Library. This "first volume" is actually Volume II; Volume I on mechanics appeared in 1826.

Similarly, in 1825, John Farrar's *Treatise on Mechanics* was published, the first of a series of "Cambridge Course" science volumes containing his lectures from Harvard;[22] and Denison Olmsted in 1831 published the first of his two-volume college natural philosophy, comprising the lectures in mechanics he had probably been giving at Yale since 1825.[23]

These three works, with all their deficiencies, were nevertheless major accomplishments. In their pages, American students were first introduced to the scientific achievements of the nineteenth century and first made aware of the variety of scientific experiments. Farrar and Olmsted utilized algebraic and geometric derivations extensively in their mechanics and even made sparing use of the calculus. Olmsted's *Natural Philosophy* was the first to be extensively illustrated and to contain many numerical problems, both as examples and as exercises.

Although the works of either Farrar or Olmsted were used in almost every college for twenty years, they were far inferior to the mathematics and astronomy texts of the same period. It was Olmsted's intention that his book "be recited after the manner of Euclid," a style of instruction rapidly disappearing.[24] This required an emphasis on rhetoric and style which too often neglected new and important physical techniques. For instance, even the 1845 edition of Olmsted's overwhelmingly popular *Natural Philosophy* contains no mathematical expressions for the laws of electricity and magnetism and, more incredibly, contains nothing of the then well-established connections between electricity and magnetism.[25]

However, other sciences were more accurately represented in their textbooks than was physics. The natural philosophy professors of the 1830–1860 period were the first American

---

[22] John Farrar, *An Elementary Treatise on Mechanics . . . designed for the use of the Student at the University at Cambridge, New England* (Cambridge, 1825).

[23] Denison Olmsted, *An Introduction to Natural Philosophy: designed as a Text Book, for the Use of the Students in Yale College* (New Haven, 1831).

[24] MS, Denison Olmsted to Prof. [Romeo] Elton, April 29, 1830, Rare Book and MS Collection, Brown University Library.

[25] Beginning with the fifth edition (1845) of some six hundred pages, Olmsted's two-volume work was printed in one and contained the entire natural philosophy course, exclusive of astronomy. After the complete posthumous revisions made by Ebenezer Snell of Amherst College (first published in 1860), Olmsted's *Natural Philosophy* bore little resemblance to the 1831 edition. However, after 1850, there were many other good books available and Olmsted's popularity dwindled.

professors to know considerably more than appeared in the texts. They were often well aware of the textbooks' deficiencies, especially in electricity and magnetism, and thus usually taught these sections of their course with lectures prepared in the library from journals and European texts. Of the course at Brown in 1833, it was said:

> With respect [to] the course of study of natural philosophy I find the Cambridge course [Farrar's] is still used though in a different manner from that first produced. Only the parts in Astronomy & Mechanics are used as textbooks & these by means of an analysis prepared by Prof. Caswell. The other parts of the course are reduced to lectures and thus presented. They commence the study the first term of the junior year, devoting one lesson a day, till the spring term of the senior year, or rather the end of that term.[26]

In addition to lectures, a few European textbooks on mechanics were available after 1833. The *Mechanics* of Boucharlat or Young gave the interested student an opportunity to pursue his studies further.[27] In fact, they became required texts at Union, Columbia, Wesleyan, and the University of Pennsylvania by 1840, although other colleges did not change for yet another decade.[28] Jonathan Pearson, while still a student at Union College, had the opportunity to use Farrar in his junior year (1833) and Boucharlat in his senior year (1834). Comparing them, he wrote,

> Bocharlat's [*sic*] Mechanics is the most pleasing work I have studied this many a day on account of the clarity of his demonstrations in which a complete application of Analyt. Geo. & Calculus is made, fully showing the superiority of his work over Farrar's. The latter is deficient in order, system & connexion in his arrangement, clearness & generality in his

[26] MS, J. Kingsbury to George W. Keeley, March 22, 1833, Brown University Archives.

[27] M. Boucharlat, *An Elementary Treatise on Mechanics* (New York, 1833), translated by Edward H. Courtenay [U.S. Military Academy]. John Radford Young, *The Elements of Mechanics* (Philadelphia, 1834).

[28] For a few schools, I was unable to find what mechanics texts were used at specific times. At Princeton, for instance, no textbook is listed in the catalog; yet it is clear from notes dictated by Joseph Henry that the level of work was higher than in the common textbooks.

> demonstrations which in attempting to make as far as possible independent of mathematics, he has materially injured.[29]

A student at Amherst, where Olmsted was still used, wrote in his diary:

> Read some of Biot . . . on electricity today. The more I read of other authors on these subjects the less I think of Olmstead [*sic*], whose only merit seems to be that of a not-over-judicious compiler.[30]

In the 1840's, pressure finally mounted to import, print, or translate more advanced European physics textbooks. Edward Hitchcock of Amherst, dismayed that so many professors were still keeping silent on the issue, publicly admitted "that many of those [textbooks] in science, are far inferior to what they ought to be."[31]

Optics, both geometrical and physical, was one area of physics which received radically changed and now adequate attention. No sooner had Enfield disappeared, with his discussions little advanced beyond Newton's *Optiks,* than three textbooks appeared to introduce the student to the wave theory and nineteenth-century physical optics. The first, and for two decades the most popular, was Alexander Dallas Bache's 1833 edition of Brewster's *Optics*.[32] To Brewster's original, Bache added almost ninety pages of mathematical derivations and problems in the analysis of reflection and refraction which he had developed at the University of Pennsylvania. He intended the work for "those of our colleges in which considerable extension is given to the course of Natural Philosophy."[33] William Bartlett at the Military Academy and Isaac Jackson at Union College also published optics texts that eventually supplanted Bache.[34] But Bache was quickly

---

[29] MS, Jonathan Pearson, Diary, Sept. 23, 1834, Union College Archives.

[30] William Gardiner Hammond, *Remembrance of Amherst* . . . (New York, 1946), p.219 [Jan. 31, 1848].

[31] Edward Hitchcock, *The American Academic System Defended* (Amherst, 1845), p.20. Hitchcock's remarks were also directed at the textbooks used in the academies.

[32] Sir David Brewster, *A Treatise on Optics* (first American edition, revised and extended by Alexander Dallas Bache, Philadelphia, 1833).

[33] *Ibid.*, title page.

[34] William H. C. Bartlett, *An Elementary Treatise on Optics* (New York, 1839); and Isaac W. Jackson, *An Elementary Treatise on Optics* (New York, 1846).

and widely received so that by 1836 almost every school had a separate textbook on optics, its first printed explanations of the undulatory hypothesis and polarization, and its first applications of the calculus to geometrical optics.

Electricity, however, did not receive its deserved share of attention in any textbook before 1850. The general physics texts prior to that time barely managed to convey the impression that there was an intimate connection between electricity and magnetism. It is clear, however, that these subjects were taught principally by lecture; for in the 1840's even the general public was well aware of the great practical application to which the electromagnet had been put in the invention of the telegraph; and every professor knew both of Henry's researches and of the worldwide excitement over magnetic variations in the late thirties.[35] As early as 1826, one professor wrote to another, "What an extensive field is opened by the Science of Electro magnetism!!!"[36] And beginning in the early thirties, colleges acquired a large selection of electrical equipment—chain of cups (voltaic pile), electrometers, electromagnets, motors, etc. The electromagnet which Joseph Henry built for Parker Cleaveland to use in his lectures at Bowdoin weighed sixty pounds and was capable of lifting a ton.

Meteorology is another example of a subject known in the early nineteenth century but not actively encouraged as an area for study until the 1830's. It was at first little more than the science of a few phenomena thought related to "meteors," but soon included everything related to climate-temperature and heat distribution, pressure and pressure variations, rainfall, storms, etc. The Americans' role in the development of meteorology was "perhaps one of greater significance than their contributions to any other field of science in the first half

[35] William Bartlett at West Point, for instance, unsuccessfully urged Joseph Henry to prepare a textbook on electricity and magnetism in 1844; documented in Wiener, *op. cit.*, p.55.

[36] MS, James F. Dana to Parker Cleaveland, Jan. 13, 1826, Dartmouth College Library. Dana, who unfortunately died the following year at the age of thirty-four, taught chemistry at Dartmouth. On a trip to Europe in 1815 to buy chemical apparatus for Harvard, he learned of the electrical research of Humphrey Davy. Just prior to his death, Dana was experimenting in electromagnetism and is known to have explained the principle of the electromagnet to Samuel F. B. Morse while he was in New York. See Dana folder, Dartmouth College Archives.

of the nineteenth century."[37] In the colleges, interest in meteorology began to increase in the early thirties. At Brown, Professor Caswell went so far as to prepare seven or eight lectures on the subject from the pages of *Silliman's Journal* and the first *British Association Report* (1832), thinking they "may some time or other be convenient for use."[38]

That time came more quickly than probably even Caswell expected; for a storm in 1836 served to precipitate a lively American debate on the nature of such phenomena. The colleges which had not yet indulged in the well-established procedure of keeping "meteorological journals" quickly began. By 1840 every college was using its newly purchased precision barometers and maximum and minimum thermometers to keep these weather records. But there was yet more substance added to the academic discipline than just the recording of data, in which students themselves often participated. Professor Albert Hopkins at Williams assigned his students the problem of estimating the height of a nearby mountain solely by thermometric and barometric measurements and their corrections—a problem that required two pages of algebra for the student to solve.[39] At Middlebury, perhaps the last school to encourage the study of meteorology, the entire faculty requested Professor Twining to add "Meteorological Observations" to the work of the college.[40] Between 1835 and 1845, contemporary meteorological discussion entered every college's curriculum.

Philosophical apparatus, as physics equipment was then called, had been a part of the college's resources since early in the eighteenth century. Few in number, these demonstration pieces were used by the professor to illustrate select principles of physics; and each college was proud to boast of its early novel static electrical generator and the like. However, there was no systematic pattern to dictate which few pieces would be imported or, more likely, privately donated to the college.

---

[37] Nathan Reingold, ed., *Science in Nineteenth-Century America: A Documentary History* (New York, 1964), p.92. See pp.92–94, 128–129, and *passim* for progress in meteorology.

[38] MS, Alexis Caswell, "Syllabus of Lectures on Barometer, Meteorology, Sound, etc." March, 1835, Brown University Archives.

[39] MS, A. Stanley Kellogg, Sept. 1, 1846, Williams College Archives.

[40] MS, Faculty Records, March 25, 1842, Office of the Registrar, Middlebury College.

Unwilling to hire professors of science on the same standing as those of other subjects, colleges were certainly not eager to spend much of their severely limited funds on equipment. Benjamin Silliman later lamented the meager store of physics apparatus owned by Yale when he was a student in the 1790's—"an air pump, a whirling table, a telescope or two, a quadrant, a theodolite, an electrical machine, the mechanical powers [probably pully, lever, screw, etc.], a spouting fountain & a few miscellaneous articles. The air pump [he added] would with difficulty kill a mouse, when placed upon its plate under a receiver, so imperfect was its exhaustion."[41] At Columbia, James Renwick reported that "when he took charge of the apparatus [1820] he did not find a single instrument adapted to the purposes of observation. Both of the barometers were broken and the only thermometer was not calculated for the purpose [of keeping a meteorological record]."[42]

With the appointment of the Swiss Ferdinand Rudolf Hassler as professor of mathematics and natural philosophy in 1810, Union College was one of the first schools to enlist the services of a man trained in science. At Hassler's request, Union immediately ordered $200 worth of equipment from Massachusetts instrument maker, John Prince.[43] In 1816 Union established a standing committee of trustees and a professor "to examine Library, Philosophical & Chemical Apparatus."[44] A committee on the "philosophical apparatus" alone had been formed by 1830, preceding identical developments at the other schools. Colleges, which had not taken stock of their apparatus for decades, suddenly found it necessary to keep a close account of what was ordered and what was broken. In 1826 Bowdoin recognized that "the philosophical apparatus was obtained about the time that the college went into operation or during the succeeding 6 or 7 years. Very little has been added during the past 10 or 15 years."[45]

Nothing better illustrates new developments in the colle-

---

[41] MS, Benjamin Silliman, Address to Alumni, July 22, 1842, Stokes Autograph Collection, Yale University Library.

[42] MS, James Renwick to Trustees of Columbia College, March 3, 1821, Columbia College Papers, Columbia University Library.

[43] MS, Minutes of the Trustees of Union College, Feb. 24, 1810.

[44] MS, Minutes of the Trustees of Union College, July 24, 1816.

[45] MS, Report of the Visiting Committee, Sept. 5, 1826, Bowdoin College Library.

giate curriculum than the increase in physics equipment which began in the late 1820's. After alumni groups were established, fund drives organized, and private donations secured, a major part of the now enlarged expense account was used to purchase scientific equipment. We can literally turn dusty pages through a decade of early nineteenth-century corporation records without noting a single expenditure on science. Yet, sometime after 1828, we usually find an initial appropriation of a few thousand dollars or more to rectify the omissions of the past and a subsequent annual appropriation to sustain this new scientific activity.

When the University of Pennsylvania took stock of its apparatus in 1828, it found it "to correspond with the Inventory prepared in 1816." The trustees immediately authorized Alexander Dallas Bache to spend $1,000 to equip the laboratory and then added $100 for the annual salary of an assistant to help keep the new equipment in order.[46] Soon Bache was enjoying annual apparatus appropriations of $250. The pattern was the same elsewhere. Not a college failed to make its first major appropriation by 1837. Middlebury, in that year the last to establish regular appropriations for science, noted that its then obsolete equipment had been largely obtained when the college opened, with a few more articles added in 1817.[47]

Most colleges, in their new-found enthusiasm for science, sent either a science professor, the president, or an agent to Europe to buy apparatus. Since the colleges were also actively increasing the size of their libraries, it was advantageous to send someone familiar with the European book trade, for the price of apparatus was well established. Wesleyan University, which opened in 1831, had spent $6,000 on science by 1836.[48] Between 1832 and 1837, Princeton purchased almost $2,000 worth of physics apparatus through Joseph Henry.[49] A good approximation is that the average college owned perhaps $100

---

[46] MS, Minutes of the Trustees of the University of Pennsylvania, November 4, 1828; January 6, 1829; October 6, 1829.

[47] William Fowler in address written in 1837, which appears in the appendix to *Addresses and Proceedings of the Semi-Centennial Celebration of Middlebury College* (Middlebury, 1850), pp.169–179. (This information from p.177.)

[48] See p.46 and footnote 111.

[49] MS, Minutes of the Trustees of the College of New Jersey, June 1844, Princeton University Library.

worth of physics apparatus in 1820, none of which was suitable for precise measurements, and owned at least twenty times that amount in 1850, including delicate instruments to measure numerous mechanical, optical, and electrical properties.

It is no wonder that space for both work and storage had to be found for all this equipment. Starting with cabinet cases and progressing to whole rooms fitted as laboratories, many colleges by 1840 had completed or were thinking of constructing whole buildings for all the sciences. Even $2,000 could buy a considerable amount of apparatus, and we are fortunate to have catalogs or lists from different periods, some of which include the price of each article. At Brown, the 1858 catalog lists about 250 items which were arranged in nine upright cases.[50] Wesleyan's catalog of 1861 lists apparatus worth $7,394.20, of which about $2,500 was spent on physics proper.[51] Ebenezer Snell, who was careful to enumerate everything including a fifty-cent "box of bird feathers" for optics experiments, needed 109 pages to describe the 500 articles in Amherst's physics collections.[52] The purchases in physics at Columbia College in 1832 numbered sixty-two articles totaling $2,018, appropriated as follows:[53]

---

[50] MS, Alexis Caswell, "List of Apparatus belonging to the Department of Natural Philosophy & Astronomy in Brown University," Sept.1, 1858, Brown University Archives.

[51] MS, John Johnston, "List of Apparatus in Wes. University," June 18, 1861, Wesleyan University Archives.

[52] MS, E. S. Snell, "Catalogue of the Articles belonging to the Cabinet of Philosophical Apparatus in Amherst College, 1852," Amherst College Library.

[53] Among the most expensive are pyrometer ($48), dipping needle ($60), declination apparatus ($100), [magnetic] azimuth compass ($50), electrometer ($100), Henry's induction apparatus ($30), collision apparatus ($50), Atwood's machine ($170), cycloidal motion machine ($48), apparatus for discharge of liquids ($90), model of hydraulic engine ($100), working model of steam engine ($150), friction apparatus ($60), heliostat ($50), Newton's refraction apparatus ($50), Fresnel's polarization apparatus ($80), microscope ($100), large camera obscura ($40), and Biot's magic lantern ($60).

MS, James Renwick, "List and Estimate of Articles required to complete the Philosophical Apparatus of Columbia College," Sept. 20, 1832, Columbia University Archives.

The phrase "practical mechanics" does not pertain to objects of commercial importance. In the textbooks, theoretical or rational mechanics refers to that performed by logic and mathematics alone. Physical or practical mechanics means the application of these principles to real objects—Archimedes screw, Hero's engine, steam engine—whether they are practical or not.

Heat	158.00
Magnetism	210.00
Electro-Magnetism	145.00
Sound	41.00
Optics	513.00
Theoretical Mechanics	501.00
Practical Mechanics	450.00
	$2,018.00

Although physics experiments can and often did merely amuse students, it is obvious that the expensive equipment was not purchased for this object. A good Atwood's Machine, for instance, cost between $150 and $200, and was purchased by every college. Yet it can be used for one and only one undramatic purpose—to measure the acceleration due to gravity.[54]

Modern apparatus, visual aids, and the more extensive use of lectures were the primary components of the student's new confrontation with science after 1830. Although lecturing was known in the college even in the eighteenth century, it was principally confined to natural philosophy classes. Eventually, there were also lectures on intellectual and moral philosophy, political economy, and government until its techniques embraced all of the newer subjects and some of the old. However, it seemed to many that lecturing conflicted with the proper "disciplining" of the mind. If the student merely sat back and took notes, how were instructors to know that he was actually participating in the learning process? Science professors were well aware of these arguments and fully sympathetic to them. Although they lectured more than the other professors, they continued to use the recitation as the fundamental system of examination.

The act itself of reciting, however, changed considerably. Instead of being asked to memorize and deliver definitions to his teacher in accordance with a recitation method used since the beginning of the American college, the student might now be called to the blackboard to derive an equation. He might be asked to "recite" from prepared diagrams or to explain what

[54] E.g., Columbia's Atwood apparatus cost $170 in 1832. Amherst's was obtained for $180 from Pixii of Paris in 1831.

the professor was doing in an experiment. Since examinations *en masse* were yet reluctantly applied, recitation remained the name for any individual performance.[55] Parker Cleaveland at Bowdoin explained the recitation-room philosophy in his first lecture to new classes:

> Something additional to textbook every recitation—which becomes a part of next lesson. . . . Study the subject as well as the textbook. . . . No textbook embraced all the facts. Some must be communicated. . . . Learn to think . . . . Imagine what questions may arise and prepare answers. . . . Additional matter from other authors—or improvements-Periodicals. Expts—App[aratus] to be engaged a part of evening . . . All a part of the lesson next morning. . . . At liberty to prepare any useful question . . . Easier to understand than to remember . . . Memory not sufficient by itself . . . Everything (notes included) to be studied, unless directions to omit. No books at recitations—I furnish the figures on an enlarged scale. Not all called upon a each recitation—but each make every question his own. . . Benefit of reciting—Lessons appear short—They require minute attention. Add[itional] matter & expts—all the hour.[56]

At Columbia, the physics schedule was as follows:

> With the Professor of Nat. & Exp. Phil. & Chemistry the Class studied one hour daily for five days each week. . . . They were taught chiefly by lectures & recitations illustrated by diagrams, apparatus and experiments, each student being required to take brief notes in the lecture room and write the subject out fully afterwards; also to submit said notes frequently for inspection and correction—all copying being chiefly forbidden, whether from the notes of another or from printed books, except in case of absence. . . .[57]

---

[55] I certainly do not claim that students did not memorize derivations; but I still believe we can sense a real change here.

[56] MS, Parker Cleaveland, "Instructions to Class" [n.d.; about 1840], Cleaveland Papers, Bowdoin College Library.

[57] MS, [Richard S. McCulloh] in Minutes of the Trustees of Columbia College, Oct. 22, 1855. The method of checking notebooks was a standard procedure and probably accounts for the fine copies now available to the historian. There are, for instance, many sets of notes at Princeton from the lectures of Joseph Henry, who also checked notebooks. A typical one contains 75 lectures, over 100,000 words, and almost 300 neatly sketched drawings and diagrams. (MS, Henry C. Cameron, "Joseph Henry Lectures in Natural Philosophy . . . 1846-7," Princeton University Library.)

Some innovations in physics teaching, such as Cleaveland's requirement that students use the laboratory during their spare time to perform experiments, were not universal. However, at all colleges certain changes were obvious after 1830. First, the student came to college with an introduction to science, however poor, which he received at the academy or common school. Second, every college exhibited an increase in time allotted to the study of natural philosophy. At some, beginning physics moved from the senior to the sophomore year, forcing the average student to study mechanics three times in his academic career. Third, extensive apparatus and numerous experiments were an integral part of the course. Fourth, a new emphasis was given to the different divisions of physics. Electricity, for instance, became a major discipline. Fifth, in the course of thirty years, the physics textbooks came to regard mathematics rather than rhetoric as its organizational principle.

In 1851, a few years after college textbooks arranged in the present mathematical manner began to appear, the *North American Review* published a thirty-seven-page review of these books and the new methods of physics instruction.[58] The *Review* echoed the dominant philosophy of the "advantage of instruction from the textbook, illustrated by oral lectures, over the exclusive adoption of either of these methods of instruction."[59] It also admitted that "perhaps there is no subject in which such lectures are a more valuable auxiliary than in physics."[60] But it was gratified that there were now physics textbooks which utilized "algebraical and infinitesimal analysis" and in which one could find "acoustics, magnetism, electricity (mechanical, chemical, animal, magnetic—and thermo—), the geometrical and physical properties of light polarized and unpolarized, optical instruments, thermotics, including the statical and dynamical laws of heat as applied to combined and radiant caloric, and photography."[61]

To suggest that after college an individual student could not apply Ohm's Law in a new situation, or could not remem-

[58] "Elementary Works on Physical Science," *North American Rev.* 72 (1851): pp.358–395.

[59] *Ibid.*, p.364.

[60] *Ibid.*, p.364.

[61] *Ibid.*, pp.362, 375.

ber how to solve an elementary differential equation in mechanics, merely focuses upon an inescapable feature of liberal education. Neither, in fact, did very many college students retain much Latin, which they studied more than physics. The fact remains that physics was now studied at greater length, taught with greater skill, and made a pillar in the shaky structure that was the ante-bellum college. We had reached the day, according to the *Review,* in which few people would "deny that scientific culture and, as one of its elements, instruction in physical science, should hold a place, and not an inconsiderable one, in every system of education."[62]

[62] *Ibid.,* p.358.

V. Astronomy

THE earliest program at Harvard College, dated 1642, included astronomy, which seniors studied for half an hour a week.[1] Although hints of the Copernican solar system were contained in New England almanacs as early as 1659, it was not until 1686, however, in Morton's *Compendium Physicae,* that Harvard students got a systematic dose of the newer astronomy. At Yale, students were probably indoctrinated in a geocentric universe until the 1720's when the works of Newton and the Enlightenment guaranteed unswerving devotion to the newer heliocentric cosmology, in addition to increasing general interest in astronomy. By the late eighteenth century, astronomy in the college had advanced beyond the stage of mere definitions to include an acquaintance with nautical tables and geographical measurement. Small telescopes of less than three-inch objective were often located at the colonial colleges, where they were used to furnish observations possible only from America's unique position on the globe. As early as 1769, public interest in astronomy was generated by our participation in the international effort of cooperation and competition to ascertain the dimensions of the solar system by observations on the transit of Venus.[2] America possessed some good observers; contributed a few improvements to the colimating telescope and Hadley's quadrant; and, like the more advanced countries, demonstrated obvious utilitarian needs for astronomical observations to be used in navigating and surveying.

---

[1] Edwin C . Broome, *A Historical and Critical Discussion of College Admission Requirements* (New York, 1903), p.22, gives the complete roster. Theodore Hornberger, *Scientific Thought in the American Colleges, 1638–1800* (Austin, 1945), p.23, tallies that astronomy occupied 2 out of a total of 140 college hours.

[2] See Brooke Hindle, *The Pursuit of Science in Revolutionary America, 1735–1789* (Chapel Hill, 1956), pp.146-165, *passim*. Also, Harry Woolf, *The Transits of Venus: A Study of Eighteenth Century Science* (Princeton, 1959).

Yet, collegiate astronomy was subject to the same periodic fluctuations as the other sciences. When especially talented scientist-teachers such as John Winthrop were not serving on a college's faculty, its students had little to compose their studies beyond poor textbooks. "Science promoter" Ezra Stiles, president of Yale, reported in 1788 that the total scientific curriculum at that school was contained in Martin's *Philosophical Grammar,* which devoted only a quarter of its pages to astronomy.[3] Labeled "Cosmology," the astronomical section of the book contained a general discussion of the universe, sun, moon, planets, comets, and fixed stars without the use of any mathematics whatsoever.[4] How far astronomy, as taught in the American college, was removed from the detailed discussions of Newton can best be illustrated by an example of Martin's rhetorical display:

> Then you suppose the *Moon* to be inhabited, I find? I have heard there be some of that opinion. Yes. Undoubtedly; to what End else can serve the Distribution of Land and Water . . . ?[5]

At the beginning of the nineteenth century—and, in fact, even as late as 1820—instruction followed the same pattern as it had in 1760. Most colleges had a set of small globes on which terrestrial and celestial indentifications could be made; but the reading of the globes was a sterile activity when never accompanied by any of the mathematical and instrumental means that allowed one actually to determine his own position on the sphere of the earth. Still worse, early in the nineteenth century, even at Harvard, the globes were so worn and neglected that "several classes have been either partially or totally deprived of that portion of their instructions."[6] Al-

---

[3] Franklin Bowditch Dexter, ed., *The Literary Diary of Ezra Stiles* (New York, 1901) 3: p.312, *passim* indicates that Martin was dropped on March 27, 1788, after 28 years.

Benjamin Martin, *The Philosophical Grammar: Being a View of the Present State of Experimental Physiology, or Natural Philosophy* (6th ed., London, 1762). No American edition of Martin was produced although it was the usual textbook, circa 1760–1790. I have only seen the sixth London edition, which references cite; the first edition appeared in 1735.

[4] *Ibid.,* pp.128–173.

[5] *Ibid.,* p.148.

[6] MS, Parker Cleaveland [tutor] to Corporation of Harvard College, Dec. 17, 1804, College Papers, Harvard College Archives.

though Princeton and the University of Pennsylvania also owned Rittenhouse orreries, by which the mathematical relations of the planetary orbits were mechanically reproduced, there is no evidence that students at these schools ever learned to calculate orbits. Moreover, not one college used a textbook which reflected any of the new continental directions given to physical astronomy. Only after the revolutionary European advances in grinding achromatic lenses and machining finely divided circles multiplied the possibilities for purely observational astronomy, were Americans forced to evaluate their system of instruction. However, a change in teaching astronomy required not only a new set of textbooks, but also a new set of instruments—instruments which American industry could not produce for decades to come and which would thus have to be imported from Europe at great expense, taxing the scientific appropriations of the colleges more than any single item had yet done. But the early part of the century was still not marked by any astronomical excitement for Americans, who felt that "no one of the natural sciences appeared to afford less hope of further discoveries than astronomy."[7]

The most widely used textbook of the early nineteenth century was Enfield's *Institutes of Natural Philosophy* which, first coming into use in the 1790's, remained the exclusive scientific text at most colleges well into the 1820's.[8] The astronomical part of the book is the longest section; but the whole astronomical portion, including appendix on lunar and solar tables, only represents the state of the science as of 1750. An American reviewer, lamenting the fact that yet a third American edition of Enfield had appeared in 1820, complained that Dolland's improvements upon the telescope (1758), "the greatest discovery in optics since Newton," were still left out.[9] In addition, known errors in Newton's *Principia,* such as the incorrect masses of Jupiter and Saturn, were preserved because

---

[7] Denison Olmsted, *An Oration on the Progressive State of the Present Age* (New haven, 1827), p.8.

[8] William Enfield, *Institutes of Natural Philosophy* [London, 1783, 1799] first American ed., Boston, 1802). The second American edition was revised from the notes of Samuel Webber of Harvard, posthumously, in 1811; it differed from the first in its inclusion of an astronomical appendix containing explanations of the use of lunar and solar tables for navigation. The third and, I believe, last American edition was published in 1820.

[9] "Remarks on Dr. Enfield's *Institutes of Natural Philosophy*—Third American Edition—1820," *Amer. Jour. Science* 3 (1821): p.146.

Enfield, obviously, had merely sat down with old astronomy books in one hand and his pen in the other. In fact, his astronomy was copied mainly from the *Principia* or the works of Rutherforth, written in the mid-eighteenth century.[10] Enfield not only preserved Rutherforth's helpful little discussions about why the moon must be inhabited, but even added proofs that the sun must be inhabited—all for the American student in the first quarter of the nineteenth century to study as astronomy! The 1821 reviewer, obviously perturbed that American students often remained unaware of the progress that was responsible for the erection of observatories in every country of Europe while America still remained without one, intoned,

> We had hoped to see the numerical expressions for the principal magnitudes, distances, and angles which occur in Astronomy, corrected in the present edition so as to correspond with the present state of the science. But . . . the numbers of former editions, many of which are a century old, are retained.[11]

Unfortunately, Enfield, the catch-all textbook, was usually taught by a catch-all natural philosophy professor who had little or no experience with modern astronomical calculations, while the low level of mathematical instruction in the colleges virtually precluded any possibility of a student's familiarity with the work of Lagrange, Laplace, and Bessel. The great charm of seeing God's handiwork manifest in the creation of the heavens, commonly cited as justification for the study of astronomy, could not hope to produce competent astronomers. Once the great number of European observatories pointed their noses to the sky, the days of the amateur professor were of necessity numbered.

The change to a system of modern observational astronomy and the introduction of physical astronomy into the colleges required a few years more than the revision of the mathemat-

---

[10] Thomas Rutherforth (1712–1771), *A System of Natural Philosophy* (Cambridge, 1748).

[11] "Remarks on Dr. Enfield's *Institutes* . . . ," p.146. The reviewer (and I) are concerned here not simply with the truth value of the statement that the sun is inhabited but rather with the context in which such a conclusion is presented. I am indebted to Deborah Warner of the Smithsonian Institution for the knowledge that even so skilled a contemporary English astronomer as William Herschel (1738–1822) could believe the sun inhabitable without making his astronomy depend upon it.

ics curriculum. As late as 1821, even the library of Bowdoin College contained as its most recent astronomical volume an eighteenth-century text reprinted in 1803;[12] and prior to 1822 only a few books circulated to make students, or more probably professors themselves, aware that Enfield's astronomy was a century old. One of these works—based upon the late eighteenth-century lectures of John Ewing, provost of the University of Pennsylvania and assistant to Rittenhouse in 1769—was published in 1809 by Robert Patterson, the Pennsylvania professor of mathematics who had communicated with Astronomer Royal Nevil Maskelyne on lunar observations for a new nautical almanac.[13] The second was the American edition of Ferguson's *Astronomy*[14] also written by Patterson, who included the recent observational triumphs of the English and gave instructions for setting a telescope into the meridian, as well as for other needed operational procedures.[15] The most popular recent English work, however, was the American edition of Vince's *Astronomy* of 1811 which, like the other texts, first appeared in Philadelphia.[16] Vince was an improvement over Enfield as were the other works; but, being an English text, it followed an exclusively geometrical line of reasoning which made it very difficult for students to use. Although its numerical quantities and observational data were more recent, the continental improvements were not even mentioned. Most discouraging was the absence of any instruction in the use of instruments, and the basing of calculations upon the use of tables rather than upon the direct reduction of astronomical data.

Yet despite the circulation of these somewhat improved volumes, Enfield continued to be used in the colleges. Each year into the new century, America grew further removed from

[12] *Catalogue of the Library of Bowdoin College in Brunswick* (Brunswick, 1821), pp.3-4. The book cited was the fourth edition of Bonnycastle's *Introduction to Astronomy*.

[13] John Ewing, *A Plain Elementary and Practical System of Natural Experimental Philosophy; including Astronomy and Chronology*, revised by Robert Patterson (Philadelphia, 1809).

[14] James Ferguson, *Astronomy Explained Upon Sir Isaac Newton's Principles and Made Easy to those who have not studied mathematics* . . . (first American ed. by Robert Patterson, Philadelphia, 1806).

[15] Patterson, in 1808, also wrote his own textbook on astronomy, which had very limited circulation.

[16] Samuel Vince, *The Elements of Astronomy: designed for the use of Students in the University* (first American ed., Philadelphia, 1811).

the forefront of the science of astronomy. In 1818, even the conservative *North American Review* complained that "We are almost the only nation of any pretensions to learning and the arts, which has totally neglected to provide for this branch of knowledge. Amidst all our public seminaries and public societies . . . we cannot boast of a single foundation for an Observatory in the United States." Lamenting the inferior position in which lack of observational tools had placed us, the *Review* complained that "We could not even calculate an almanac without help from abroad."[17]

The first astronomy with all the European advances, descriptions of modern apparatus, and instruction in the manipulation of astronomical data was written by John Gummere in 1822.[18] A member of the American Philosophical Society (1814), Gummere ran a private boarding school from which he accumulated enough money to buy his own observational instruments; and in 1833 he became first professor of mathematics at Haverford College, where he soon erected a small observatory.[19] Gummere's book was followed in 1827 by the astronomy of John Farrar of Harvard College;[20] and soon after, John Herschel's *Astronomy* was imported to America in its 1833 London edition only to be prepared by Sears Walker for its first American edition in 1836.[21] Two more works which appeared in 1839—one by Denison Olmsted of Yale and the other by William Norton, who later also taught at Yale—gave the colleges a choice of five textbooks for use in the era.[22] Benjamin Pierce apparently gave his Harvard stu-

[17] *North Amer. Rev.* 8 (1818): p.200.

[18] John Gummere, *An Elementary Treatise on Astronomy* (Philadelphia, 1822). Further editions by the author appeared in 1837, 1842, and 1846. The editions of 1851, 1853, 1857, and 1861 were revised by E. O. Kendall of the University of Pennsylvania.

[19] Willis I. Milham, *Early American Observatories,* pp.48–49.

[20] John Farrar, *An Elementary Treatise on Astronomy, adapted to the Present Improved State of the Science, being the Fourth Part of A Course of Natural Philosophy, compiled for the use of the Students of the University at Cambridge, New England* (Cambridge, 1827; second ed., 1834).

[21] John Herschel, *A Treatise on Astronomy*. A new edition by S. C. Walker (Philadelphia, 1836).

[22] Denison Olmsted, *An Introduction to Astronomy: designed as a Text Book for the Students of Yale College* (New York, 1839). This work was revised by the author through the sixth edition of 1846. It was then continued and rendered far more readable for students in the revisions of Ebenezer Snell of Amherst College, which appeared from 1861 to 1883.

William A. Norton (1810–1883), *An Elementary Treatise on Astronomy* (New York, 1839). Further editions appeared in 1845, 1854, 1867, 1881.

dents a choice of Farrar, Olmsted, or Herschel.[23] But of greatest importance is the fact that there now existed responsible textbooks of astronomy and that by 1833 Enfield's *Institutes* had finally disappeared.

Despite the great proliferation of popular astronomies and textbooks on mathematics and other sciences, these five astronomies alone served all the colleges for decades. Herschel's *Treatise,* the most widely used of all, was still assigned at the end of the century; and its last American edition was published in 1902. In both size and content, these five astronomical texts were far superior to those in use during the entire previous century. Older texts were as short as a hundred pages or less, with whole passages wasted in the nonsense of rhetorical display or in the redundancies of mathematical imprecision; these new books were composed of entire volumes of at least three hundred pages, in which can be found detailed descriptions of the mounting, setting, and use of all astronomical instruments, most of which the colleges did not even own in 1820. But the most revolutionary feature of the new texts were the uses to which they expected a student to put his observations. The telescope was no longer considered an instrument of magnification for the sake of mere beauty alone; books were no longer concerned with quadrants and tables of navigation; precession, aberration, and nutation were not glossed over in a paragraph. Rather, the student was expected, in principle, to incorporate all known celestial anomalies with his own data from modern scopes, circles, and clocks in precisely describing a body's location or motion. "Mathematical investigations of the fundamental principles of the science . . . [replaced] facts and descriptions.[24]

The inspection of any of these textbooks, however, will not reveal the entire content of any single course in astronomy, since, in the absence of professors of astronomy *per se,* the subject was taught by colleges' professors of natural philosophy or mathematics. Physical astronomy or celestial mechanics—the application of mathematical analyses to the general problems of gravitation—was, for instance, taught at almost every college, often constituting a distinct group of lectures

math more imp't

[23] Harvard, *Catalogue (1845–46),* p.44.

[24] MS, John Farrar to [President and Fellows of Harvard College], Dec. 26, 1825, College Papers (XI), Harvard College Archives.

entitled "central forces"; yet this study came not from the textbooks of astronomy, but rather from those of integral calculus or of analytical mechanics.

In this now sophisticated age, when a second of longitude or time was considered beyond tolerable error and a good astronomical clock did not fault by more than a few tenths of a second a day, it became necessary to secure firm foundations to mount instruments. The kind of information yielded by portable telescopes and the kind of locating possible with nautical instruments no longer had any relevance to the calculations of the astronomer. Therefore, the erection of permanent observatories to house college telescopes, circles, and clocks, was the unique development of decade 1830–1840.[25]

The typical astronomical equipment of the college in 1828 consisted of one or more portable telescopes, a sidereal clock, and sometimes a portable transit instrument—none of which, of course, could be used for precise determinations. In 1822 when Professor James Renwick wished to determine the longitude of New York from atop the cupola of Columbia College with the college's new Dolland telescope and box chronometer, he could not rely upon his own instruments. For help, he called upon a New York watchmaker who, like others in Boston and Philadelphia, had erected an "observatory" fitted with a permanent transit instrument and astronomical clock.[26] William Cranch Bond, Boston watchmaker, erected at his home in the early 1820's a similar private observatory, which by 1839 had grown to the Harvard College Observatory. But despite periodic proposals for a national observatory and some small activity by college professors for a decade or so, Astronomer Royal George B. Airy could truthfully remark in 1832 that there was not "any public Observatory in America, though there are some able observers."[27]

Erection of observatories 1830-40

---

[25] The dating of observatories of all kinds is rendered difficult by one's definition of "observatory." It is not my intention to describe each college observatory, its instruments, their date of procurement, date of mounting, etc. The interested reader is referred to Willis I. Milham, *Early American Observatories: Which was the First Astronomical Observatory in America* (Williamstown, Mass., 1938), which contains a bibliography, pp.53–58, on early observatories.

[26] "Longitude of New York Ascertained by the observations of Prof. James Renwick" *Amer. Jour. of Science* 5 (1822): pp.143–144.

[27] G. B. Airy, "Report on the Progress of Astronomy during the present Century," *Report of the British Association* (at Oxford, 1832), p.130.

Events of the 1830's, however, soon fostered interest in astronomy on many levels. The meteor showers of 1833 provided a spectacle for both the American public and the American astronomers who were able to identify the Leonids as celestial bodies;[28] and the knowledge that there would be worldwide competition when Halley's comet returned in 1835 spurred American enthusiasm for achievement in astronomy. Moreover, at the beginning of the decade, modern astronomy had finally become part of the college curriculum; and new sources of financial aid, principally the philanthropy of many private donors, gave the colleges more money to spend. When every college sent either an agent or an officer to Europe during the decade to buy apparatus, astronomical equipment was usually included in the list of purchases.

The University of North Carolina had $6,000 to spend on books and equipment in 1824. It ordered from Europe a meridian transit telescope, altitude and azimuth telescope, and astronomical clock with mercurial pendulum which were finally mounted in a building completed late in 1831. Meanwhile, Yale had received $1,200 from a single donor in 1828 to import what was then the largest telescope in America—a Dolland refractor with five-inch objective.[29] But the makeshift mounting of this telescope in a building constructed for another purpose eventually diminished its usefulness as a research instrument; it was hauled about a floor in a wooden carriage without a firm equatorial mounting and could only sight objects thirty degrees above the celestial horizon. However, Professor Denison Olmsted and tutor Elias Loomis used the telescope to sight Halley's comet on August 31, 1835, weeks before the news of its appearance arrived from Europe.

Olmsted's success with the new Yale telescope contributed to the excitement over the erection of observatories that had begun in the colleges; but it was not the sole source of enthusiasm. Observatories had been discussed by educators for a decade, although it was only in the 1830's that money for

---

[28] The co-discoverers were Denison Olmsted (1791–1859), professor at Yale College (1825–1859), and Alexander Catlin Twining (1801–1884), then studying at West Point and later professor at Middlebury College (1839–1849). See *Amer. Jour. Science* **25** (1834): pp.363–411 and **26** (1834): pp.132–174 and pp.320–352 for the presentation of their results.

[29] David F Musto, "Yale Astronomy in the Nineteenth Century," *Ventures* (Spring, 1968), pp.7–17, contains interesting suggestions about the support of astronomy throughout the century.

their establishment could be found. But whether an individual observatory was the result of personal zeal or collective competition, there was a realization that, if astronomy was to mean more than surveying, navigation, and star-gazing, new installations were required. Tutor Loomis went to Western Reserve College (then at Hudson, Ohio) in 1836, where the observatory he constructed still stands. Albert Hopkins of Williams College sailed for Europe in the fall of 1834 to procure sidereal clocks, Troughton and Simms transit instruments, and an Herschelean reflector of ten-feet focus. In 1836 Hopkins began to quarry stone for the foundations of the Williams College Observatory, which was built with the help of the students and opened in 1838. Probably the first observatory of true permanence, with revolving dome, mounted telescope, and circles, the building itself is still in use today.

The story of college observatories, 1830-1860, is complicated by the wide variety of instruments, mountings, foundations, etc., employed in them. All were not of equal value as research centers, although they were all good enough for student instruction, which was, in any case, their main purpose. Amherst, for example, first purchased a transit instrument for $500 in 1832; yet it was not until 1846, after a $9,000-fund drive for science to which Abbott Lawrence had donated $1,000 was completed, that the college constructed an observatory. When the permanent Lawrence observatory was finished in 1848, a repeating circle and astronomical clock had been added to the transit instrument; but there was still no other telescope suitable for mounting.[30] In 1839 Harvard took over the private observatory of watchmaker William Bond.[31] But, after the dazzling comet of 1843, many citizens and learned societies of Boston contributed to a much larger observatory which, at its opening in 1847, contained a $20,000 German telescope with fifteen-inch objective, 2000 magnification, and parallactic mounting.[32]

[30]MSS, Minutes of the Trustees of Amherst College, August 21, 1832, and July 6, 1847. William B. Calhoun, *et al., Addresses at the Dedication of the New Cabinet and Observatory of Amherst College, June 28, 1848* (Amherst, 1848), p.11. For description of instruments and photograph of Amherst Observatory, see David Todd, "Early History of Astronomy at Amherst College," *Popular Astronomy,* No.106 (June, 1903), pp.1-5.

[31]MS, C. C. Felton to Francis Bowen, Dec. 27, 1839, Harvard College Archives, contains a description of the arrangement.

[32]Elias Loomis, *The Recent Progress of Astronomy; especially in the United States* (New York, 1850), pp.184-194.

Within a few years after Wesleyan opened, President Fisk and a young tutor, Harvey B. Lane, set sail for Europe to buy books and apparatus. Augustus W. Smith, professor of science at Wesleyan who was directing purchases from home, thought that a fully equipped observatory also required an altitude and azimuth instrument in addition to a meridian instrument. As it was common for Americans to seek advice at the great European observatories, Smith's collaborator took his single 'scope to the Paris Observatory, writing home of the suggestions he received there:

> Prof. Arago [director of the Paris Observatory] who is perhaps better acquainted with optical instruments than any other person . . . says that all the instruments strictly necessary [for] an Observatory are a Transit Instrument & mural circle for Meridional Observation, an Equatorial Inst. for the Observing of objects out of the Meridian, together with a clock . . . there is no altitude & azimuty inst. in the Observatory at Paris, nor, I think, in that at Greenwich. Prof. Arago says that it is not necessary to purchase another telescope for the equatorial Inst. but that the one already purchased could be adjusted . . . [and] would afford every faculty for instruction and would be sufficiently accurate for most observations.[33]

In addition to advising Lane on such matters, Arago pointed Wesleyan's 6-inch telescope upon the satellites of Jupiter to test its achromatic quality at magnifications up to 350. The University catalog of 1836–1837 announced the use of the new instruments by students; and a wooden observatory was built in 1839. Surveying the decade of astronomy, a Middlebury professor remarked in 1839,

> Till of late little has been done in this department in our land. But not only the old states are waking up to the subject, but even the west, and south-west are expending considerable sums in supplying to their infant institutions the facilities for philosophical and astronomical observation.[34]

---

[33]MS, Harvey B. Lane to Augustus M. Smith, April 29, 1836, Wesleyan University Archives.

[34]Alexander C. Twining in *Addresses delivered at the Inauguration of the Professors of Middlebury College, March 18, 1839* (Middlebury, 1839), p.43.

The Observatory of Philadelphia, which opened in 1840, started a new era in American observatories. There, for the first time, Americans became aware of the finer quality of German optics. Erected by the City of Philadelphia in connection with its new Central High School, the observatory was constructed with elaborate iron foundations and marble mountings to insure stability. Its instruments were ordered in Germany, probably at the suggestion of Alexander Dallas Bache, who toured Europe for Girard College, 1836–1838; and its 6-inch equatorial scope, ground by the celebrated Munich firm of Merz and Mahler, had variable magnifications to 480.[35]

Following Philadelphia in 1838, almost all American institutions making purchases secured at least one Merz telescope—West Point (1839), Georgetown College (1841), the long awaited National Observatory (1843), Cincinnati Observatory (1845), Harvard (1847), Amherst (1848), Dartmouth (1848), Union (1856), and Dickinson (1858).[36] By the mid-1840's, observatories were becoming an indispensable feature of college education. Most colleges, by 1850, had an observatory either already erected or in some stage of development, even if a permanent building was not finally completed until after 1860.[37] The report of William Smyth, professor of mathematics at Bowdoin, illustrates the manner and meaning of acquiring astronomical tools:

> A refracting telescope that would cost from $500 to $1000 would form a most valuable addition to our apparatus, and would serve as the nucleus for the instruments of the proposed

[35]In its early days, Central High School had very close associations with the University of Pennsylvania. Specifically, the schools used each other's scientific resources. In its first decade, the eminent High School faculty included Alexander Bache, John Fries Frazer, Sears Walker, and Ezra Otis Kendall. Walker edited the American edition of Herschel's *Astronomy* and Kendall revised Gummere's text. All but Walker were at some time associated with the University. Kendall, whose work on stars and comets at the High School Observatory secured him and the Observatory an international reputation, became professor of astronomy at Pennsylvania, 1854-1894. See Loomis, *op. cit.*, pp.165–168.

[36]The famous Dudley Observatory of Union College, opened in 1856, is detailed in Loomis, *Recent Progress* (second ed., 1856) and in *Amer. Jour. Education* 2 (1856): pp.593-604. (Article entitled "Dudley Observatory at Albany, New York.")

[37]To my knowledge, the only schools which may not have owned extensive astronomical apparatus by 1860 nor started collections for an observatory were Vermont and Middlebury.

Observatory of Philadelphia

> Observatory. On account of the want of a lecture room, and finally with reference to the procuring of such an instrument, I have reserved about $100 of the appropriation of the last year—and should an annual appropriation of $200 only be made for a few years to come, the instruments might be safely ordered. As such instruments are made only to order, and cannot be procured excepting in Europe, something like two years must probably elapse before it could be obtained. On this account it is very desirable that the initiatory steps for the purpose should be immediately taken. An instrument of the value referred to, would be sufficient for exhibiting to the classes the most interesting objects brought to view in the modern astronomy, and would be an invaluable acquisition to our means of instruction.[38]

Courses

College astronomy consisted of recitations, lectures, and instrument and observation hours. Astronomy *per se* occupied at least one full term in the junior or senior year; while analytical dynamics or mathematical astronomy was either part of a second term or part of the last course in mathematics or natural philosophy. Although most schools owned good instruments, they did not all require the same use. More than half the schools had observatories; and, of course, those that did not had to limit their field work to the older practical instruments and stargazing. Thus, at Princeton, which did not get an observatory until 1868, a student of the fifties remarked that his classmates did not have "an opportunity for celestial instrumental study beyond that of a small telescope."[39] At Yale, practical astronomy was made an optional senior elective in 1845.[40] But generally, if a school had the resources, it was proud to boast in its catalogs that the

[38]MS, William Smyth in "Report of the Visiting Committee," August 28, 1845, Bowdoin College Archives. The plan was executed.

[39]Typed MS, Alfred Alexander Woodhull [Class of 1856], "Reminiscences of Princeton," p.31. A copy of the original manuscript of 1912–1914 can be found in Princeton University Library.

[40]Yale, *Catalogue (1845–46)*, p.29. The word "practical" was much overworked in nineteenth-century America and is often ambiguous. In astronomy, it did not denote surveying operations and the like, but distinguished the purely observational astronomy from that requiring mathematical manipulation. All the textbooks include, therefore, such nonpractical operations as calculating the effect of nutation on the position of a star among "practical" operations. Edward Everett summed up the distinction: "The brilliant discoveries of astronomy are for the most part made with the first class of instruments [telescopes, alone];—its practical results wrought out by the second [precise instruments to measure angular distance]," in "The Uses of Astronomy," *Amer. Jour. Education* 2 (1856): p.608.

"Lectures in Astronomy are accompanied by celestial observations and instructions in the use of instruments."[41] Any school, after all, could take advantage of commercial sets of "drawings for the illustration of the lectures in . . . Astronomy."[42]

But whatever the instrumental resources, the study of astronomy now required more time than a single term. From a half hour a week in the seventeenth century to perhaps two or three hours at the beginning of the nineteenth century, astronomy now demanded five or more hours. Eventually, the time devoted to other topics such as optics had to be shortened in order to make "more time for astronomy which is much too extensive a science to be embraced in a single term."[43] There was enough time, however, to go through the complete textbook. A Dickinson professor reported at the conclusion of his course that the senior class had "studied Norton's Astronomy entire . . . [and] the adjustment and use of the principal Mathematical Instruments and other practical applications in the field."[44] A professor at Brown reported:

> he heard the recitation of the Senior Class in Norton's Astronomy five mornings in the week with the exception of two, in which lectures were delivered. He gave also, at other hours, two lectures extra . . . . In addition to the regular studies, he also while prosecuting the study of Astronomy with the class, devoted a portion of from twelve to fifteen evenings, occupying from 1 to 2 hours each to observations with instruments & to the study of sidereal astronomy in the open air.[45]

Caswell calculated that astronomy occupied six hours per week for fourteen weeks or eighty-four hours.[46] About eighty

---

[41] Dartmouth, *Catalogue (1849),* p.27.

[42] Princeton, *Catalogue (1838–39),* p.20. Wesleyan University uniquely displayed the world's only Russell Orrery. The outer planetary orbit of this two-thousand pound curiosity was forty-five feet in circumference. The orrery was used both by students and as a public planetarium for which admission was charged. (Wesleyan, *Catalogue (1837–38),* p.22; and printed advertisement, June 16, 1837, Wesleyan University Archives.) Many colleges and lower schools had small demonstration orreries, inperfectly scaled, and usually made of wood. They were available from American manufacturers for about $20.

[43] MS, Alexis Caswell to Francis Wayland, July, 1841, Brown University Archives.

[44] MS, Wm. H. Allen in Minutes of the Trustees of Dickinson College, July 12, 1843.

[45] MS, Alexis Caswell to the Faculty of Brown University, July 28, 1843. Brown University Archives. Reports from professors are often written in the third person.

[46] MS, Alexis Caswell to [?], July 27, 1844, Brown University Archives. I use the term hour in its literal sense and not in the modern sense of semester hour.

hours is a good estimate for the time devoted to astronomy in the era 1840–1860. Alexander Dallas Bache advised his successor at the University of Pennsylvania that students could well handle ten pages a lesson in Gummere's *Astronomy*, requiring forty hours just to get through the textbook.[47] This would leave an equal amount of time for lectures, which, it must be remembered, were never used to replace recitation but "are delivered to the Class after they have finished their recitations in the Text-book and are therefore presumed to be acquainted with the elements of Astronomy." They then formed "the basis of the examinations both public and private."[48]

Students also found encouragement for study outside of the classroom. A student at Amherst spent his "leisure hours in part with astronomical observations," borrowing a telescope from the college to use on his own, and assisting Professor Snell with his observations at the college observatory.[49] Astronomers at Harvard College and Union College also found willing assistants among the students.

But, of course, not all students cared for astronomy; and its increased sophistication gave rise to problems not even defined in 1800 or 1828. A Princeton student described the general student's despair in the astronomy course:

> A very learned man . . . he [Stephen Alexander] was not a happy teacher for the average student of those days. The wall of the lecture room facing the class was filled with enormous black-boards. These would be filled with mathematical formulae day after day and week after week the connection in which once lost was not recovered. A statement once made from his desk was not repeated. Prof. Alexander seemed to assume that whatever he had once announced was understood and remembered, which excepting by a few gifted (or very studious) men was not the case . . . . the roll was followed in such regular order that it was very easy to forsee when any particular student would be called upon and to arrange oneself accordingly.[50]

---

47 MS, Alexander Dallas Bache to John Fries Frazer, March 21, 1844, Frazer Papers, Library of the American Philosophical Society.

48 Denison Olmsted, *Outlines of A Course of Lectures on Astronomy, Addressed to the Senior Class in Yale College* (New York, 1852), p.2.

49 MS, George Gardiner Hammond Diary, July [16,], 1848, Amherst College Library.

50 Typed MS, Alfred Alexander Woodhull [Class of 1856], Princeton University Library, p.31.

We would certainly not characterize the period by pretending that all students enjoyed their mental training. The many had astronomy forced down their throats, while only the few had ample material and human resources to foster their interests. At least, however, as one astronomy professor of the nineteenth century described,

> By 1861 this country had acquired what I may call a school of astronomers. That is, there were many observatories. . . . Any young man who . . . felt himself impelled to study astronomy, could find instructors; and after sufficient training could usually get remunerative employment. He could gain the ear of a public interested in such things.[51]

Though aided by the government, astronomy was definitely oriented around the colleges, their professors, and their observatories.[52]

---

[51] Truman Henry Safford, *The Development of Astronomy in the United States* (Williamstown, Mass., 1888), p.22.

[52] As an example, the first AAAS Committee on the Prime Meridian contained professors from eight of these sixteen schools—Alexander (Princeton), Anderson (Columbia), Caswell (Brown), Frazer (Pennsylvania), Lovering (Harvard), Smyth (Bowdoin), Snell (Amherst), Stanley (Yale): *Proc. Amer. Assoc. Advancement of Science,* Second Meeting, Cambridge, August 1849 (Boston, 1850), p.viii.

VI. Chemistry and Its College Derivatives

Chemistry could probably have never gained and sustained a popularity relative to physics and astronomy if its boundaries had not been extended considerably beyond what presently comprises chemistry. In the second quarter of the century, almost every academic investigator designated a chemist also worked in the area labeled geology. In addition, other fields of natural history also had a close relationship to chemistry. Indeed, the origins, intellectual justifications, techniques, and destinies of all these disciplines were so intertwined that, if we attempt to separate them too exactly, we lose sight of the single impact which together they made upon the college structure. Chemistry and geology, for instance, were both concerned with mineral analysis far into the nineteenth century, and it is usually unclear to which discipline the science of mineralogy belonged. Chemistry and natural history both had their collegiate origins in the lectures designed for medical departments. Although for the first third of the century chemistry was distinctly more popular than botany, after 1840 their study and popularity conveniently converged on problems of agriculture. Geology and natural history were the sciences most intimately connected with the natural theological doctrine of design and were often the joint objects of those state surveys, launched in the late 1830's, whose results had such dramatic impact upon the college. Thus, it is not profitable to seek constant distinctions among these topics.

A distinction which should be made, however, is one between the history of chemistry in America and its development in the context of the college; for analysis of the advancement of a science itself often obscures educational issues. Thus, the historian of American chemistry is more concerned that the second quarter of the century saw the establishment of the nation's first private analytical laboratories than he is about the less dramatic but more subtle fate of chemical knowledge within colleges only dedicated to teaching it. He finds the appointment of a first-rate experimenter by a medical school to be of more significance than the appointment of a second-rate chemist by a college that previously had none. Yet, for our purposes, the developments he overlooks are of first importance, being indications of fundamental historical change within the college.[1]

At the beginning of the nineteenth century, chemistry was the only science in America clearly in the ascendancy. A Phi Beta Kappa orator in 1802 boasted that

> no part of science, within twenty years, has received as great improvements as chemistry. . . . Our country may already boast, of a long list of chemists, the number and accuracy of whose experiments, are not inferior to those of European chemists; and who are ardently engaged in the advancement of this science.[2]

By contrast, geology had not been heard from in America, and botany was "more neglected by our citizens, than any of the whole circle of the sciences . . . we are dependent upon the Europeans, for the knowledge of the plants of our own country."[3] But each of these sciences eventually entered the college curriculum; and as they did, their study was entrusted to professors already teaching other subjects. By 1860, five well-defined sciences (excluding mathematics) were being taught—physics, astronomy, chemistry, geology, and biology—but generally by only three professors whose titles reflect both the various combinations of subjects made in the college and the multifarious interests demonstrated by these men themselves in their private scientific lives. In brief, there were professors of geology and natural history, natural philosophy and geology, chemistry and geology, and even one of zoology and astronomy, the most frequently used title being that of professor of chemistry and natural history.

---

[1] There is no need to explore further the differences between the points of view of an historian of chemistry and one of education. The problem is basically an historiographic one, rooted in the differences between institutional history or personal biography on the one hand, and more comprehensive political, social, or intellectual history on the other. It is necessary only that the tenets and biases of the two approaches be clearly understood.

[2] Eli Ives, *An Oration on Chemistry and Botany* (New Haven, 1803), p.12.

[3] *Ibid.*, p.13.

It is well to keep in mind that the teaching of mathematics and all science by a single professor had virtually disappeared from the college by 1830 and that the most immediate pressure for this development—from which mathematics, physics, and astronomy reaped benefit—was the growth of the particular bodies of knowledge discussed in this chapter. The serious study of geology which commenced in the college in the 1830's and the study of organic chemistry, argiculture, and physiology which similarly marks the early 1840's were responsible for the intense debate from which electives, parallel courses, and scientific schools emerged. But the most immediate void filled by these special scientific programs was not advanced training in the various physical sciences, engineering, or mathematics, but the training of chemists, specifically analytical chemists, to assay minerals, soils, and organic products.

Since chemistry was the science most successfully pursued in the United States during its first half-century of independence, it is particularly important to distinguish among chemistry in general, chemistry in the medical school, and chemistry in the college.[4] That chemists in America should have so assiduously kept abreast of the most recent European chemistry while allowing physics and mathematics to languish in a decrepit state of scholarship is interesting. Even botany, which had played such a large role in the colonial awakening to science and which required no mathematics for further study, fell far behind chemistry and the progress of recent discovery. Undoubtedly, chemistry's special favor arose from its prominent position in medical education and its many practical applications in industry, or the arts, as manufacturing processes were known. Until the widespread applications of steam power and electricity supplied new attention-getting alternatives, the social betterment derived from applications of chemistry to the industrial arts defined the claims of American science upon public support.

[4] A modern study of the content of chemical courses in the colleges is Wyndham Davies Miles, *The Development of Chemical Education in Columbia, Princeton, Rutgers, and Pennsylvania* (unpublished doctoral dissertation, Harvard University, 1955). Although I am indebted to Miles for some suggestions and particularly for his analyses of individual chemistry courses over a long period, I feel that he has taken too sanguine a point of view in focusing upon the best chemists and the best teachers.

References for a comparative history of education in chemistry are in Ludwig F. Haber, *The Chemical Industry during the Nineteenth Century* (Oxford, 1958), p.264. It is interesting to note that Haber concludes that provision for chemical instruction was greater in America than in Dalton's England even in the first part of the century (p.37).

Many of the influential eighteenth-century American chemists had been educated at the University of Edinburgh, where both William Cullen and Joseph Black had emphasized that "chemistry was a practical, useful science, and not merely a philosophical pastime";[5] and a Yale student noted that Silliman extended Black's formal textbook definition of chemistry to stress its role in "the improvement of the arts."[6] Samuel L. Mitchill of Columbia College, who gained America's first collegiate appointment in chemistry, carried the awesome title of "Professor of Natural History, Chemistry, Agriculture, and the Other Arts Depending Therein."[7] Similarly, the first chemistry professor at Dartmouth was expressly required to lecture in the "application of Chemistry to the Arts"[8]—a phrase which to many was synonymous with chemistry itself. Despite the fact that pottery, glassware, and steel processes usually made their way into the college course, they never became the principal focus of study. Chemistry courses always began with discussions of heat and its effects upon matter, before entering upon other theoretical areas such as the mechanisms of electrolysis.[9]

Although there was little instruction in the crassest material considerations such as how to estimate the price of an ore, study in the recognition and analysis of minerals was a substantial part of every chemistry course, and a part that could not help but be noticed by industry. The emphasis placed upon theory by the college course, coupled with the ever widening separation of such study from the most immediate industrial concerns, is one of the reasons that chemical courses outside of the core college curriculum were eventually established. That the phrase "chemical applications to the arts" should nevertheless have so often been flaunted before

[5] Quoted in Miles, *op. cit.*, p.40. Also, see Archibald and Nan L. Clow, *The Chemical Revolution* (London, 1952).

[6] MS, Isaac Townsend, "Notes of Chemical Lectures delivered by Professor Silliman," Jan., 1825, Yale University Library.

[7] *Catalogue of the Governors, Trustees, and Officers; and of the Alumni and the Graduates of Columbia College, in the City of New York from 1754 to 1867* (New York, 1868), p.23. Appointed in 1792, Mitchill retired in 1801, and his vacant chair was not refilled; he also taught in the medical school.

[8] MS, Minutes of the Trustees of Dartmouth College, Oct., 1820.

[9] Americans, like some Europeans, were strongly committed to the actual existence of caloric as one of the "imponderable agents" of nature. Among the first performed demonstrations in a lecture course was usually a proof that caloric had no weight, accomplished by weighing a solid object at two widely varied temperatures.

the public by trustees and chemists alike is an interesting example of the means that must be employed to find support for science. Depite the rhetoric, however, so far as the college was concerned, "No teacher would be tolerated, who should more than incidentally allude to any common use, like that of economical profit, that could be made of them [chemical applications in the arts]."[10]

There can be no doubt that after 1790 chemistry caused the greatest scientific excitement in America. The new "French System" was widely embraced here within a few years after its European codification, while twenty years elapsed before the French mathematics was finally introduced. By 1810, Dalton's new atomic theory was already being taught, although college astronomy was a half-century or more behind the times.[11] Even the second American edition of Enfield's *Institutes* (1811), in which the entire natural philosophy course was then contained, omitted the section on chemistry since "more complete works in the subject have . . . become common in the country."[12] At Yale College, where Benjamin Silliman was appointed professor of chemistry and natural history, provisions were made for the teaching of six times the amount of chemistry as of natural history and botany combined in the first revision of the college laws to include these studies.[13] By 1816, a Harvard chemist was calling the period the "age of the atomick theory."[14]

Undoubtedly, chemistry in the college benefited from the work of the corps of professional chemists at the medical schools of the University of Pennsylvania, Columbia, and Harvard. The great majority of the chemistry professors

---

[10] Charles U. Shepard, *An Address . . . before the Agricultural Societies of Hampshire, Franklin, and Hampden Counties at their Anniversary Fairs* (Northampton, 1847), pp.21-22. Such an admission does not indicate that some teachers were practicing hypocrisy, for there really was a legitimate connection between laws and their operation in industrial processes. It does, however, indicate that a difference existed between the expectations of those interested in immediate application leading to an increase in productivity and profits and those concerned with knowledge, not necessarily conceptually abstract, but at least freed from the burden of cost analysis.

[11] The atomic theory became known in America, as in Europe, through Thomas Thomson, *Elements of Chemistry* (1808).

[12] William Enfield, *Institutes of Natural Philosophy* (2nd American ed., Boston, 1811), p.xii.

[13] *The Laws of Yale-College . . . enacted by the President and Fellows* (New Haven, 1808), p.16.

[14] John Gorham, *Inaugural Address* (delivered Dec. 14, 1816) (Boston, 1817), p.19.

appointed in our sixteen colleges from 1820 to 1840 received their training either at one of these three medical schools, the oldest in the country, or under the tutelage of Benjamin Silliman, who himself learned chemistry at the Medical School of the University of Pennsylvania.[15] Less well publicized is the fact that many other colleges established connections with local medical departments in the first few decades of the nineteenth century; yet by 1828, such arrangements had been made at Dartmouth, Brown, Yale, Vermont, Middlebury, Bowdoin, and Williams.[16] Thus, from the standpoint of the liberal arts curriculum alone, the teaching of chemistry in the first decades of the nineteenth century did not progress steadily. Rather, its establishment in that curriculum required a protracted struggle among medical-chemical lecturers, overworked arts faculty, and trustees, during a period in which chemical activity itself moved from the medical schools into the hands of the college's permanent officers.[17]

Benjamin Silliman reported that in 1792 Yale, which did not yet have a medical department, hardly knew chemistry "even by name."[18] John Maclean, appointed professor of chemistry and natural history at Princeton, which also had no medical school, soon found that he was really expected to teach primarily natural philosophy; and until his retirement in 1812, he constantly complained of "the disadvantages I am under here" getting the faculty to approve chemistry.[19] After seven years of failure, Brown in 1817 finally found a regular chemistry professor who was required to bring his own

[15] An interesting "family tree" for the education of scientists (principally chemists) from this period is found in Donald Beaver, *The American Scientific Community, 1800-1860: A Statistical-Historical Study* (unpublished doctoral dissertation, Yale University, 1966), p.170.

[16] A short description of most medical schools with student enrollments, etc., is in *Quarterly Register and Journal of the American Education Society* 2 (1830): pp.245-247.

[17] Catalogs often pointed out that lecturers of chemistry and professors of medical sciences were not regular members of the faculty, or "permanent officers." Miles, *op. cit.*, p.3, identifies the period 1810-1847 as that in which chemistry was established in the college, thus spanning the time from the earliest professorial appointments to the use of separate analytical laboratories. I myself think that we can identify the years 1825-1835 as the period in which firm foundations for chemistry were established in the overwhelming majority of liberal arts courses.

[18] MS, Benjamin Silliman, "Address to Alumni," July 22, 1842, Stokes Autograph Collection, Yale University Library.

[19] MSS, John Maclean to Benjamin Silliman, Feb. 1, 1804; March 8, 1804; Feb. 10, 1810, Princeton University Library.

chemical apparatus;[20] and the University of Pennsylvania, which invited Thomas Cooper to lecture in chemistry and mineralogy in the college, offered him nothing "but a room."[21] The University of Vermont offered a professor "no salary" but the opportunity to collect fees from "students & others, who shall attend his lectures;"[22] and Union College provided its first chemist with a modest salary and the chance to collect fees from non-matriculated students who might wish to take the course.[23] Even at Harvard, it was not until 1827 that a chemistry professor taught primarily in the college rather than in the medical school.

And thus it continued throughout the 1820's. A medical professor would lecture in the college for a year, returning to private practice for his moonlighting fees. A special lecturer would find that he could not collect enough money on student entrance tickets to make a living and therefore would be forced to leave his post. Similarly, in 1823 when Benjamin Silliman could not deliver his chemical lectures because of illness, he urged Parker Cleaveland to substitute by assuring him that he could come during the vacation "in the Academical part of our Institution," indicating that the medical students were still the prime target of his chemistry course.[24]

Finally, however, chemistry became a topic of concern in every college proper. Entering the arena of discussion about collegiate reform in the 1820's, it succeeded everywhere in emerging with its first group of professors as regularly paid members of the arts faculty. At Middlebury College, for instance, the following resolution passed in 1826:

> Whereas the income of the corporation is barely sufficient to meet the expense of the institution upon its present establishment with four instructors, and whereas it is desirable if not absolutely necessary to provide an additional instructor by filling without delay the chair of the professorship of Chemistry & Natural History, and whereas there is reason to hope that

[20] Donald Fleming, *Science and Technology in Providence, 1760–1914* (Providence, 1952), p.29.

[21] MS, Thomas Cooper to Parker Cleaveland, July 20, 1817, Dickinson College Library.

[22] MS, Minutes of the Trustees of the University of Vermont, July 31, 1813.

[23] MS, Minutes of the Trustees of Union College, July 23, 1811.

[24] MS, Benjamin Silliman to Parker Cleaveland, Sept. 27, 1823, Silliman Family Correspondence, Yale University Library.

a fair and full presentation of the exigencies of this case may induce some friend or friends of learning and their country to endow this professorship. Resolved that the Rev. President Bates be authorized as an agent for this corporation to make such representations to the wealthy and generous, and solicit donations for the purpose above. . . .[25]

Probably the most conservative of the colleges, Middlebury added a chemistry (and natural history) professor immediately, making the number of instructors for both mathematics and science only two. By 1836, many colleges had added two professors to their faculty and were able to extend the study of chemistry from three classes a week in part of the senior year to five classes a week in the entire junior year or more. A Columbia professor about to reintroduce chemistry "as part of the general [natural philosophy] course" assured that "this obstacle will be in some measure overcome when the instruction is extended, as it is contemplated, to two classes instead of one, and shall include two whole years of the collegiate course."[26]

As late as 1813 Harvard described itself as "destitute of the necessary implements & vessels for the [chemical] experiments";[27] and even in 1817, when Union College, one of the most scientifically progressive, was shopping for chemical glassware, it discovered that its manufacture was still "new in our country."[28] But a quarter-century later there were "artists who fabricate adroitly every article of chemical apparatus."[29] Chemicals, too, which early in the century had to be purified

[25] MS, Minutes of the Corporation of Middlebury College, December 18, 1826. A professor was appointed on Jan. 2, 1827, and given a year's leave (with salary) to prepare himself for the job.

[26] James Renwick, *Inaugural Discourse* (delivered Jan. 4, 1821) (New York, 1821), pp.19, 21.

A tabular summary of studies at twenty colleges appears in *Quarterly Register and Journal of the American Education Society* 1 (1829): pp.228–232. It illustrates colleges in many stages of transition from their having still no separate chemistry course to their teaching of almost two years of chemistry. The old and new mathematics curricula, etc., are also neatly contrasted in these pages.

[27] MS, "Report of Committee on Library" in Records of Overseers, VI, Oct. 26, 1813, Harvard College Archives.

[28] MS, Minutes of the Trustees of Union College, July 23, 1817. There is an often repeated story that early in his career at Yale Benjamin Silliman sent a broken retort to a glass manufacturer as a model and got back a shipment of similarly broken retorts. See George P. Fisher, *Life of Benjamin Silliman,* (New York, 1866) 1: p.126.

[29] *Sketches of Yale College with numerous anecdotes . . . by a member of that institution* (New York, 1843), p.93.

or derived from cruder substances, could easily be acquired in this age when colleges needed them.[30]

In addition, a gradual change was wrought in the concept of the college laboratory. At first, a "laboratory" was little more than a room in which chemicals were stored, and later it became a better ventilated room with ovens and benches. Finally, however, ample space and facilities were provided for any kind of inorganic analysis and the handling of gases. In 1836, for example, when Oliver Hubbard reached Dartmouth, one of the last schools to regularly employ chemical demonstrations, he wrote:

> But to my Laboratory—Chem Lab. it can hardly be called. The furnaces in a block of masonry are two—one for heating a *sand bath alone* & another for heating an *iron pot* (for water) 10 or 12 qts. capacity—they can be used for nothing else. A black furnace in the lecture room—but the chimney running up thro the middle of the lecture room—has been cut off in the upper story . . . & is *useless*. There is not a bell glass or any article for pneumatic Chem. No means of making oxygen. Cistern of wood . . . will not hold water—pestles without mortars . . . *one flask*, long *narrow* neck—no [test] tubes . . . reagents and chemicals decomposed . . . the utmost destitution of every useful object & only the ruins of what once was. . . .[31]

Dartmouth immediately prepared to correct this deficiency, appropriating $700 for the chemistry laboratory and constructing, like other schools, special buildings to house the physics and chemistry apparatus.[32] Because chemicals and chemical glassware had to be constantly replenished, they provided a major expense for the colleges, but one that the schools were now willing and able to bear. In contrast to physics or astronomy for which a few thousand dollars each were needed for modernization, the initial outlay to equip a chemistry laboratory was less than $1,000.[33] But interest was so

[30] For example, in 1802, John Maclean was unsure whether he could get "red precipitate of mercury [HgO, probably to repeat Priestly's experiments] in Philadelphia." MS, John Maclean to Benjamin Silliman, Feb. 1, 1802, Princeton University Library.

[31] MS, Oliver Hubbard to B. Silliman, April 23, 1836, Dartmouth College Archives.

[32] See p.40.

[33] The inventory of scientific apparatus at Wesleyan University in 1861 totals $7,300, of which about $800 was housed in the chemistry laboratory. John Johnston, "List of Apparatus in Wes. University," June 18, 1861, Wesleyan University Archives.

great that funds were appropriated for the construction of special laboratories to house these chemical wares.[34] By 1838, Harvard, Yale, Princeton, Pennsylvania, Union, Dickinson, Bowdoin, and Wesleyan even provided for laboratory assistants (often students) to keep laboratories in order, while the University of Virginia provided a black "servant . . . to aid the Professor of Chemistry."[35]

Chemicals and laboratory equipment were arranged and employed in a number of ways through the years. Early in the century, the natural philosophy professor had at his disposal a few odd pieces and chemicals which he merely displayed during a traditional recitation period, while chemistry professors were apt to rely more and more upon "experimental lectures." In his first hour lecture to chemistry classes at Bowdoin, for example, Parker Cleaveland demonstrated solution, double decomposition, and precipitation in the course of performing seven operations.[36] Students, too, sometimes handled the apparatus either when called to the counter for a recitation, when attending specially assigned laboratory hours, or when given the permission of the professor. By 1850, every student had considerable exposure to the laboratory and had seen or performed numerous qualitative and quantitative analyses in addition to experiments designed to corroborate the prevailing theories of atomicity and combination. An 1854 list of laboratory equipment at Brown contains hundreds of articles of glassware, common and special; balances, "a complete set of volumetric appartus," and "23 sets of Reagent bottles for student use containing 32 [chemicals] in each set."[37]

During the period 1825–1835 when chemistry made its most dramatic gains in the college, a new class of young men was teaching the subject. Unlike their predecessors who had taught the full circle of science with the frequent complaint

[34] A detailed floor plan of the Harvard College laboratory with a list and location index of all the articles soon found in all college laboratories can be seen in John W. Webster, *A Manual of Chemistry* (Boston, 1826), preface, vii; but only in the first (1826) and second (1829) editions.

[35] MS, Minutes of the Rector and Visitors, July 20, 1827, University of Virginia Archives.

[36] MS, A. S. Packard Jr., "Prof. Parker Cleaveland's *Lectures in Chemistry . . . 1855,*" Bowdoin College Library.

[37] "List of Apparatus in the Laboratory. Sept 1, 1854," [unsigned], Brown University Archives.

since 1810 and 1815 that chemistry had "passed out of their field of view," these men had more time and opportunity to devote to chemistry.[38] They quickly extended the scope of recitation and introduced laboratory experiences; and virtually all of them adopted as a text Turner's *Chemistry* (London, 1827), which was used through many editions until about 1850 when kinetic-molecular concepts appeared in newer textbooks.[39] Turner's *Chemistry* was divided into four parts—I. Imponderables, II. Inorganic, III. Organic, IV. Analytical; and Alexander D. Bache at the University of Pennsylvania succeeded in taking his class "thru nitrogen and its compounds" or most of the book.[40] At newly reorganized Brown, a faculty report in 1830 relates how the new courses, taught with the new text, were conducted:

> Two Thirds, at least, of this work [Turner] including all the most important parts of it, were studied with care, and subjected to frequent and thorough revisions. The recitations were uniformly held in the laboratory, and accompanied with experimental illustrations, twice or thrice a week, as the subject might demand. It must be obvious to every one that this method is calculated to put the student in far more complete and permanent possession of the principles of the science, than

[38] Denison Olmsted, "On the Present State of Chemical Science," *Amer. Jour. Science,* **11** (1826): pp. 349–358 and continued in **12** (1827): pp.1–14. This quotation from **11**: p.349.

[39] Edward Turner, *Elements of Chemistry,* American editions prepared by B. F. Bache of the Franklin Institute in Philadelphia (Philadelphia, 1828–1840). Replaced by Wesleyan Professor John Johnston's *A Manual of Chemistry, on the Basis of Dr. Turner's Elements of Chemistry* (Philadelphia, 1840).

I do not analyze the pre-1828 textbooks such as Dana, Gorham, Kames, Chaptal, or Brande because, unlike the textbooks of mathematics, physics, and astronomy, they illustrate the expected progression in the state of the science. Many editions of both Bache's and Johnston's editions of Turner have a twenty-page appendix, added to the book or bound separately, containing questions for students. In the appendix to this study is a sample of these printed questions, owned by a student at the University of Pennsylvania in 1841 and now housed in the Edgar Fahs Smith Collection of the University Library. Chemistry lectures, unlike those of physics, very closely follow the format of the textbooks. See, for example, syllabi of Joel Nott (1825), Union College Library; Benjamin Silliman (1825+), Yale University Library; or Parker Cleaveland (various years), Bowdoin College Library.

[40] MS, Alexander Dallas Bache to John Fries Frazer, March 21, 1844, Frazer Papers, American Philosophical Society. Eventually the first section of chemistry on imponderable agents (caloric, light, electricity, galvanism) was removed or shortened as it was absorbed into physics courses and more organic chemistry was provided at the end of the course.

texts

> that of merely hearing lectures, which, till within a few years, has been in such general use.[41]

As in physics, the student was often required to write out his lecture notes and present them to the professor for inspection. A student at Bowdoin even found that the second year of chemistry was "much more interesting to me than last year: I take notes, 'copiously'—It takes much time to write them out but I think it well spent."[42]

Mineralogy and geology, both unknown in the college or even in the country in 1800, were soon slowly added to the "chemistry" course until, in about 1835, separate semesters of mineralogy and geology were common. In reply to a question about the inclusion of mineralogy in the chemistry course, Benjamin Silliman advised another professor on its importance:

> The reasons, why I thought you should "by all means" attend to mineralogy, are such in my view perfectly decisive. There is so intimate a connexion between chemistry and mineralogy, that no person can sustain a respectable mark as a chemist without being able to decide at sight, or with chemical examination, on the nature of any mineral substance, which may be presented to him. The greater part of the subjects of chemical experiment and examination are from the mineral kingdom, and a lecturer on chemistry could neither retain, nor even acquire the confidence of his pupils, without a competent knowledge, I mean a considerable knowledge of mineralogy.[43]

Mineralogy, like chemistry, came to be identified and justified to the public mind with specific economic motives, although the rapid advancement of this science wrought by professionals followed a completely different course. Parker Cleaveland, largely responsible for bringing mineralogy to the attention of American scientists with the first American

---

41 Annual Report of the Faculty" in appendix to *Catalogue of the Officers and Students of Brown University* (Providence, 1830), p.21.

42 MS, James William Hanscom, "Diary," March 18, 1846, Bowdoin College Library.

43 Benjamin Silliman to Gamiliel Olds, printed in Gamiliel S. Olds, *Statements of Facts, Relative to the Appointment of the Author to the Office of Professor of Chemistry, in Middlebury College and the termination of his connexion with that College* (Greenfield, Mass., 1818), p.15.

book on this subject, initially became interested in minerals when an industrial concern asked him to identify a substance it had stumbled upon in 1805. Benjamin Silliman performed mineral analyses for both government and individuals; and James Renwick was invited to Georgia in 1833 to look for the gold in the Cherokee lands. A business community which found mineralogy "one of the most trifling and useless of all human pursuits"[44] in 1820 was clamoring for professional advice in 1840 to avoid indiscreet purchases and wasted mineral excavations.

Since mineralogy was incorporated as a chapter into most chemistry books used in the early years of the nineteenth century, it is difficult to say exactly when the subject was first taught in the course. Frederick Hall, professor of mathematics and natural philosophy at Middlebury College, reported in 1810 that "mineralogy is not comprised in my department of instruction" although he himself was studying the subject.[45] Indeed, as late as 1817, Harvard College still did not own a compound blowpipe, a primary tool of mineralogical analysis.[46] However, with the chemistry course as short or nonexistent as it was in the first years of the century, exactly when mineralogy was universally required cannot be that important a consideration. When Cleaveland's *Mineralogy* appeared in 1816, the *North American Review* used the occasion to remark that we "still find, with two or three eminent exceptions [probably Bowdoin and Yale] that our public means of instruction in mineralogy are extremely limited," adding that the problem was not so much one of books as of "interesting and well informed instructors."[47] A decade later, the *Review* noted that "Since the publication of Professor Cleaveland's work, mineralogy has advanced with rapid strides" and has become "a branch of study in so many of our schools and higher seminaries."[48] By the 1830's, two separate works on mineralogy appeared—one by C. U. Shepard and the other by

Mineralogy texts

[44] Frederick Hall, *An Oration, on the Importance of Cultivating the Sciences . . .* (Baltimore, 1828), p.16.

[45] MS, Frederick Hall to Benjamin Silliman, November 12, 1810, Silliman Family Collection, Yale University Library.

[46] MS, College Records, 5, May 3, 1817, Harvard College Archives.

[47] "Cleaveland's Mineralogy," *North Amer. Rev.* 5 (Sept., 1817): p.412.

[48] "Comstock's Mineralogy," *North Amer. Rev.* 15 (April, 1827): pp.487-488.

James D. Dana; and one or the other became the required textbook at every college.[49]

Both Shepard and Dana typified the new class of young scientists who were growing up in the age of *Silliman's Journal.* A significant group had interests which ranged over chemistry, mineralogy, geology, and natural history; and they were able to exercise these interests both in the many kinds of surveys which were being made across the land and in the newly created professorships of chemistry and natural history which awaited them in the colleges. By 1836, the surveys served as both a source and a resource of college professors, teaching them their skills and providing summer employment.[50]

Totally unknown in the college of 1800, geology was another of the topics which at first slowly entered the chemistry course and later received independent encouragement. Even by 1820, most colleges did not make any provision at all for the teaching of geology, while in a small number of schools it was the subject of a few lessons attached to the end of the few weeks' course in mineralogy near the conclusion of the all-inclusive chemistry course.[51] Meanwhile, a small but "tolerably skillful set of [American] nature examiners" were making "constant contributions to the knowledge of the

[49] Charles U. Shepard, *Treatise on Mineralogy* (New Haven, 1832; second part, 1835). James D. Dana, *System of Mineralogy* (New Haven, 1837).

[50] I believe it unprofitable to make a sharp chemistry-geology-biology distinction among these professors. Although many were specialists in a particular area of science, their greatest amount of time in the college was devoted to chemistry. Among those who served as principles or assistants in state geological and botanical surveys or in national exploring expeditions between 1820 and 1840, and who also taught college for at least five years were: C. B. Adams (Middlebury, 1839); Farrand N. Benedict (Vermont, 1833), James Dwight Dana (Yale, 1850); Ebenezer Emmons (Williams, 1833); John F. Frazer (Pennsylvania, 1844); Edward Hitchcock (Amherst, 1825); Denison Olmsted (Yale, 1825); Henry D. Rogers (Pennsylvania, 1836); Charles U. Shepard (Amherst, 1847); Zadock Thompson (Vermont, 1851); John Torrey (Princeton, 1830). If we add to this list those who, without regular appointment, nevertheless acted as private consultants, worked on canal surveys, reported on local geological surveys, or conducted chemical analyses for the surveys, we will have mentioned most chemistry (etc.) professors prior to 1860.

[51] I cannot possibly hope to discuss how geology itself was changing prior to study in the college. For some contemporary sense of the impact made on geology by the use of fossil evidence and the hypothesis of gradual change, see W. D. Conybeare, "Report on the Progress, Actual State, and Ulterior Prospects of Geological Science," *Report of the British Association* (at Oxford, 1832), (London, 1835), pp.365–414. This volume also contains reports on other sciences.

Shepard & Dana

Geology

geologists and naturalists of Europe."[52] After 1820 they found financial support from the western exploring parties to which they attached themselves, and eventually from individual states, which authorized partial surveys of the mineral resources of their lands. Of course, while doing this "surveying," these young geologists were using the evidence they uncovered to support favorite theories of geological change. Edward Hitchcock, first president of the Association of American Geologists, regarded the "state geological surveys . . . as peculiarly American" and praised this "favor of government" in making geology popular.[53]

In the late 1820's, American interest increasingly focused on the role of fossils in determing the history of the earth and by the middle thirties turned to more sophisticated problems of dynamical geology. During this decade (1830–1840), a few more college professors introduced the new discipline into their chemistry lectures. At first, in the absence of a textbook, professors used only a fifty-page introductory chapter at the end of Cleaveland's *Mineralogy,* which Edward Hitchcock of Amherst, one of the earliest geologists and teachers of the discipline, asked Cleaveland to publish separately, since in 1828 Amherst planned to require geology of all students while leaving parts of mineralogy optional.[54] But before Cleaveland had a chance to extend his geology or even to publish it in its present form, Benjamin Silliman prepared the first geological textbook intended for college students: an American edition of

Silliman's Text

A more recent evaluation of early American geology is Leonard G. Wilson, "The Emergence of Geology as a Science in the United States," *Jour. World History* **10** (1967): pp.416–437.

In addition to general histories of geology, the growth of the science in America is contained in the works of George P. Merill, "Contributions to the History of American Geology," *Annual Report of the Smithsonian Institution (1904)* (Washington, 1906), pp.189–733; *Contributions to a History of American State Geological and Natural History Surveys: Bulletin of the United States National Museum* **109** (Washington, 1920); and *The First One Hundred Years of American Geology* (New Haven, 1924).

[52] MS, Benjamin Vaughan to Parker Cleaveland, April 7, 1823, Benjamin Vaughan Papers, American Philosophical Society.

[53] Edward Hitchcock, *First Anniversary Address before the Association of American Geologists, at their Second Annual Meeting in Philadelphia, April 5, 1841* (New Haven, 1841), pp.5, 43.

[54] MS, Edward Hitchcock to Parker Cleaveland, November 21, 1827, Cleaveland Papers, Bowdoin College Library.

Bakewell's *Geology*,[55] which so impressed pioneer American naturalist Amos Eaton that he promised to recommend it "wherever I go."[56]

Bakewell's *Geology* was adopted at a number of schools; and by the early 1830's, geology, which in the 1820's had been taught at four or five schools, had become entrenched in the college curriculum. The process of geology's assimilation, now rapid, was accentuated in 1835. In that year, Hitchcock published the first complete state survey of Massachusetts; surveys of a half-dozen other states were initiated; and Lyell's monumental *Principles of Geology* reached America. Edward Everett, with his customary volubility, hailed Hitchcock's work as awakening "thousands of liberal minds in every part of the Commonwealth . . . to the intelligent observation of nature."[57] The *North American Review* observed that by the "aid of geology we are directed to our great mineral resources, which constitute an important part of the solid basis of national prosperity in the useful arts."[58]

Within five years, all colleges which had not yet provided for a separate course in geology did so. The first membership list of the Association of American Geologists (1840), forerunner of the American Association for the Advancement of Science (1848), contained professors from twelve of the sixteen schools then open;[59] and Hitchcock's *Elementary Geology* of 1840, an elaboration on his own lectures at Amherst, became the colleges' principal textbook, going through thirty-one editions by 1860.[60]

Natural history and natural theology, from which botany, zoology, anatomy, and physiology emerged, were the last of

Surveys
Lyell

Nat'l hist

[55] Robert Bakewell, *An Introduction to Geology* (first American ed. by Benjamin Silliman, New Haven, 1829). The appendix to this work (128 pages) contains the lectures which Silliman gave at Yale during the 1820's.

[56] MS, Eaton to Benjamin Silliman, July 28, 1828, Williams College Library. Eaton himself receives limited treatment in this volume because his primary affiliation was with Rennselaer Institute rather than with one of the liberal arts colleges; his views on education, however, are interesting and may be found in Ethel M. McAllister, *Amos Eaton: Scientist and Educator, 1776–1842* (Philadelphia, 1941).

[57] Edward Everett, *Address delivered before the Literary Societies of Amherst College* [August 25, 1835] (Boston, 1835), p.20.

[58] "Professor Hitchcock's Report on the Geology, &c. of Massachusetts," *North Amer. Rev.* 42 (April, 1835): p.423.

[59] Edward Hitchcock, *First Anniversary Address . . . ,* pp.46-48.

[60] Edward Hitchcock, *Elementary Geology* (New York, 1840).

the major scientific disciplines to find acceptance in the college. Despite a few scattered lectures on botany or anatomy, it was not until 1840–1845 that these subjects received serious attention. They were the last straw added to the camel's back before alternatives to the prescribed college course became realities.[61]

Natural history, as it was known to its important American practitioners in the early nineteenth century, found little support in the college or the community. When naturalists were bringing forth "all curious bugs, flowers, seeds, etc. . . . to receive a name, most others stood idly by."[62]

> Many a speculator on the capital bequeathed by philosophy, readily admits the utility of chemistry, mechanics, and astronomy; but meets the naturalist with "Cui Bono?" What! to chase a butterfly with the ardor of an urchin.[63]

In 1835 the *North American Review* complained that natural history was still the most neglected area of science and advocated its study "not because it may lead to something useful, but because, without looking to any other result, the mere pursuit is improvement to the mind and heart." But the *Review* did not hesitate to list all the possible practical benefits—"new uses for animals, vegetable and minerals, and ways in which they can be applied to the benefit of man."[64] These kinds of practical results, which had to be recognized by the college before any great enthusiasm for natural history could be aroused, were not, however, to become obvious until after 1840.

practical results necessary

Therefore, until that time, the study of natural history in the college was focused in two areas, subject, of course, to the

---

[61] William M. and Mable S. C. Smallwood, *Natural History and the American Mind* (New York, 1941) must be read with great caution. It has little to say about the post-1830 period because that is when "science begins." It is "inclined to suggest that the attempts of Hitchcock [in reconciling religion and geology] . . . and the position taken by Silliman really necessitate their being designated as naturalists, as contrasted with the true scientists, of whom Thomas Cooper was one of our earliest and best examples" (p.247). Unfortunately, this attitude is of very little help in understanding science in the college where Hitchcock and Silliman, and many others like them, exerted a profound influence.

[62] MS, John Torrey to Amos Eaton, April 29, 1820, Brown University Library.

[63] William H. Allen, *An Address before the Cuvierian Society of the Wesleyan University* [July 31, 1838] (New York, 1838), p.4.

[64] "Study of Natural History," *North Amer. Rev.* 41 (Oct., 1835): pp.417, 425.

abilities of individual instructors—in courses of natural theology and in student-centered natural history clubs. Prior to the general use of separate textbooks on botany, zoology, anatomy, and physiology in the 1840's, these subjects were studied from two books—Paley's *Natural Theology* and Smellie's *Philosophy of Natural History*.[65] Both had as their object the demonstration of the great chain of being, the moral superiority of man, and the necessity for the design of every biological object in its present form. Of the eight *Bridgewater Treatises* which tried to extend the doctrine of design to all of the sciences, only Roget's *Physiology* was regularly used in the American college.[66] The judgments of Paley, Smellie, or Roget on the overall plan of nature and their discernment of this plan in the organs of all living things was not a disguise for lack of factual knowledge. All of the organs, systems, life cycles, and ecologies were studied just as minutely and scientifically as if the "design" did not exist. As celebrated a naturalist as Dickinson's Spencer F. Baird used Paley's *Natural Theology* and Agassiz's *Principle of Zoology* simultaneously.[67] The concept of cosmic purposefulness represented in natural theology was not a dogma that set limits to biological inquiry. Much like twentieth-century America's belief that science could only thrive in a democratic-capitalistic system, its gradual erosion marked no transition in the way that scientists carried out their specific tasks.

The lagging support of biological science is not so much reflected in poor textbooks as in the inadequate study time accorded it. In 1835 the time given to biology was less than that given chemistry, geology, astronomy, or even optics. As Parker Cleaveland explained in 1840, "to judge of the expediency" of increasing biology, you need to know more than the

---

[65] William Paley, *Natural Theology: or the Evidences of the Existence and Attributes of the Deity* (London, 1802; with many editions throughout the nineteenth century). William Smellie, *The Philosophy of Natural History* (first American ed. prepared by Dr. John Ware, Boston, 1824; also with many editions).

[66] Peter Mark Roget, *Animal and Vegetable Physiology Considered with Reference to Natural Theology* (second ed., London, 1834), This is the fifth Bridgewater Treatise "On the Power, Wisdom, and Goodness of God as Manifested in the Creation."

[67] MS, S. F. Baird, "Report to the President, 1849," Dickinson College Library. Baird, one of most influential nineteenth-century American naturalists left Dickinson in 1850 for a distinguished career at the Smithsonian Institution (1850-1888), becoming the second secretary after Joseph Henry's death in 1878.

"mere mention of the textbook employed . . . the Time devoted to this study is only between 2 and 3 weeks. . . . Very little use is made of Smellie except the Introduction . . . [which] contains a condensed view of Cuvier's arrangement of the animal kingdom. This view is extended by means of the Lectures."[68]

Two or three weeks was not, as everyone recognized, very much time in which to study any subject. However, between 1825 and 1840 students at many colleges established natural history clubs, which became less active after the curriculum broadened. The purpose of these clubs was to collect objects of natural history, which were later absorbed by the college's own collections. Students employed the same methods of acquisition as were common among other naturalists; they swapped duplicates with other student societies and solicited individuals and lyceums to make donations. Undoubtedly, the students were encouraged in these activities by the widespread lyceum movement, whose first expressed purpose was to provide "collections in natural history."[69]

Beginning with Yale's purchase of the celebrated Gibbs Mineral Cabinet in 1825, the colleges themselves all began to acquire mineral cabinets and, later, fossil cabinets and other cabinets of natural history. The colleges purchased or received these collections from private naturalists, sometimes were legally sanctioned to act as repositories for specimens gathered in state surveys, and even executed a few imaginative schemes, like requesting students to bring "various minerals found in the various sections of the country in which they reside."[70] Princeton was so bold as to print in the catalog that "Specimens of plants, minerals, animals, shells or of ancient coins for the different cabinets belonging to the College will be thankfully received, and also any preparations suitable for the Anatomical Museum."[71]

To a large extent, the quality and quantity of individual collections were determined by the various donors. Thus,

---

[68] MS, Parker Cleaveland in "Report of Visiting Committee," August 21, 1840, Bowdoin College Archives.

[69] *American Lyceum, with the Proceedings of the Convention held in New York, May 4, 1831 to Organize the National Department of the Institution* (Boston, 1831), p.6.

[70] MS, Minutes of the Trustees of Union College, July 21, 1831.

[71] Princeton, *Catalogue* (1832–1833), p.13.

Middlebury and Amherst had a large group of C. B. Adams's mollusks, while Dickinson had many of S. F. Baird's birds. These cabinets of natural history became more and more elaborate until, in the 1840's, some schools required special rooms or buildings to house them. The public was usually invited to view these museums and thus rendered more willing to provide for their expansion. Haphazard collection was supplemented by planned purchases, especially of "new minerals and fossils and geological drawings as may appear necessary to sustain the reputation and usefulness of the Institution."[72] Whereas in 1819 Harvard had to be "roused from its long neglect of . . . [geology, etc.] to cause its few but valuable specimens to be properly arranged . . . ," by 1840 every school boasted that these subjects were "illustrated by a well selected and extensive cabinet."[73]

But although by 1840 there was little complaint that the year or more devoted to the study of chemistry or the semester or more accorded geology was not enough for the general student, the same cannot be said for natural history. A student speaking at Wesleyan's commencement in 1846 used all of the faculty's traditional ideology to ask:

> Has not a science which affords so many advantages for disciplining the mind, that creates a tone of truth and morality, that inspires the soul with a reverence and love of the Deity, the strongest claim on the attention of every friend of Education? Should it not have a prominent place in all our literary institutions, and not be dismissed by a few dry lectures, as a subject of little importance. Yet how few faculties are afforded, how meagre the libraries on this subject, how small the cabinets with specimens thrown together . . . . They are kept rather to be exhibited on great occasions as the magnificent gift of some noble hearted patron of the institution than to promote the science of Nat. Hist.[74]

The colleges answered these questions in the same way that they had answered demands associated with chemistry a

[72] MS, Minutes of the Trustees of Yale College, Aug. 19, 1845.

[73] "Systems of Geology," *North Amer. Rev.* **8** (March, 1819): p.414. Williams College, *Catalogue* (1838), p.18.

[74] MS, Caleb P. Wickersham, "English Oration: Claims of Natural History," August 5, 1846, Wesleyan University Archives.

quarter of a century earlier. The records of the 1840's are once again riddled with the names of medical doctors hired to give lectures on the biological sciences until finally, about 1850, regular professorial appointments became common and full semesters of biology were taught. Yet just at the time that physiology and anatomy were becoming popular subjects, dramatic new achievements were breaking down the intellectual barriers which had so long separated chemistry and natural history. In 1816 Harvard's John Gorham had echoed the then tenable position that "the departments of the chemist and of the physiologist are well defined. Living matter is the property of the latter, inert inanimate nature of the former."[75] But, by 1840, the growth of organic chemistry and specifically the work of Justus Liebig in demonstrating that "the chemical forces play a part in all the processes of the living animal organism," caused the chemists to add a strikingly increased amount of organic chemistry to their courses.[76] This startling realization had an impact even beyond the chemistry course, for organic chemistry merged into the biological courses and accelerated interest in chemical agriculture. Chemistry already owned the lion's share of the scientific studies in the liberal arts curriculum. Could the college hope to add life sciences, organic chemistry, organic analysis, and agriculture besides?

Certainly, agriculture and societies for its promotion were not a new phenomenon, since even in the eighteenth century these local associations had flourished to disseminate empirically derived improvements in scientific agriculture. Nor was the nineteenth century ignorant of the use of fertilizer, even organic fertilizer; Davy's *Agricultural Chemistry* was widely circulated in America and rendered readable to any literate farmer,[77] while there was abundant literature on "how the elements of the soil may be determined by analysis."[78]

---

[75] John Gorham, *Inaugural Address* [December 14, 1816] (Boston, 1817), p.21.

[76] Justus Liebig, *Organic Chemistry in its Applications to Agriculture and Physiology* (second American ed. by John W. Webster (Cambridge, 1840), preface, p.xix.

For the reception of Liebig's work in America see Margaret Rossiter, "Justus Liebig and the Americans: A Study in the Transit of Science, 1840–1880" (unpublished doctoral dissertation, Yale University, 1971).

[77] Humphry Davy, *Elements of Agricultural Chemistry* (Philadelphia, 1821).

[78] William James Macneven, *Introductory Discourse to a few Lectures on the Applications of Chemistry to Agriculture, delivered before the New York Athenaeum in the Winter of 1825* (New York, 1825), p.19.

Athenaeums, lyceums, and special schools like the Gardiner Lyceum in Maine (1823) propagated the gospel of soil content, and the subject also received a few paragraphs in college chemistry textbooks. But Liebig's work established the necessity of analyzing the plant's total ecology including the atmosphere and the character of the soil in which it grew. He was not the first to demonstrate that nitrogen fertilizers were efficacious, for every farmer knew that by 1840; but he showed the importance of the balance of the nitrogen cycle. The many analytical experiments conducted on plants changed the chemist's outlook if not yet the farmer's.

Professor Charles B. Adams of Middlebury called for farmers to "secure the services of an agricultural chemist in every state."[79] William Seward begged the farmer to be governed not by "rules" and "magical influences" but by the "laws of Nature."[80] Professor Hitchcock of Amherst asked "what scientific improvement can the farmer make, who understands no [chemistry and geology]?"[81] And Professor Charles U. Shepard of Amherst remarked that "the farmer is by no means singular in his inacquaintances with chemistry, natural history, and physiology, though from the nature of his occupation he suffers more than others by such ignorance."[82] But older agrarian attitudes were rapidly disappearing in the 1840's, and by 1850 Professor Ebenezer Emmons of Williams College could insist that "we now hear very little complaint [about books on scientific agriculture which] ten years ago, would be excluded from the fireside of the farmer because [they] employed names and words difficult to be understood."[83]

How was this new importance for scientific agriculture to affect the college? "Many persons appear to think, that our college course can be so modified as to fulfill at the same time, the literary and the agricultural requisition." But to Professor

---

[79] Charles B. Adams in *Addresses delivered at the Inauguration of the Professors of Middlebury College, March 18, 1839* (Middlebury, 1839), p.22.

[80] William H. Seward, *The Elements of Empire in America* (New York, 1844), p.26.

[81] Edward Hitchcock, *American Academic System Defended* (Amherst, 1845), p.22.

[82] Charles Upham Shepard, *An Address before the Agricultural Societies of Hampshire, Franklin, and Hampden Counties at their Anniversary Fairs* (Northampton, Mass., 1847), p.6.

[83] Ebenezer Emmons, *Agriculture of New York* (Albany, 1851) 3: preface.

Shepard "it does not appear . . . that such a plan is likely to succeed."[84]

Of all the new demands that were placed upon the college in the 1840's, the liberal arts curriculum was prepared to meet only one of them—namely, the inclusion of biology in the course of study. In 1847, at the same time that Shepard was moving about the Massachusetts countryside delivering his speech on agriculture, Amherst appointed a professor of zoology; and from 1840 to 1850 most colleges expanded in these areas. In addition, organic chemistry and biology combined earned themselves another semester in the liberal arts program. The lecture techniques used for two decades in the teaching of the other sciences became the medium of student-teacher exchange in the instruction of these new ones. Just after organic chemistry and vegetable physiology were added to the lectures at Brown for the first time, a student wrote, "How a fellow can make up on these I don't exactly comprehend."[85] But textbooks on all of these subjects quickly supplied the solution. In addition to the obvious visual aid of the specimens in the "cabinets," charts, diagrams, and "exhibition[s] of drawings of animals by magic lantern" were employed.[86]

By 1850, there were usually four professors of mathematics and science on faculties which numbered about nine. The University of Vermont, which did not get its fourth scientist until 1854, received a letter of resignation from an overworked professor in 1847:

> Forty years ago, scarcely anything was done in connexion with our American College on the subject of either chemistry or Natural History,—very little even thirty years ago. But within that time a great change has arisen. Those branches have

---

[84] Shepard, *An Address before the Agricultural Societies,* p.20.

[85] MS, G. Newell to Eli Thayer, June 8, 1843, Brown University Archives. Translation: How do I get a set of notes when most of the material is not in a textbook?

[86] MS, John L. Lee, "Notes on Edward Hitchcock's Natural Theology Lectures," Sept., 1843, Amherst College Library. The magic lantern was a one candlepower "slide projector." Glass plates either etched by professors or purchased commercially provided biological drawings. Every school owned a lantern, and such exhibition of slides in class seems to have been its principal use. For a picture and explanation, see Denison Olmsted, *An Introduction to Natural Philosophy* (New York, 1845), p.569.

> formed so high an importance both in their scientific & practical relations that all the colleges with which I am acquainted, except this one, have found it needful or expedient to make large appropriations for them both, & have found a great extension of their influence & usefulness from so doing.[87]

Biological topics growing out of the course of chemistry were the last subjects to be introduced in the required curriculum; this extension of science into the liberal arts course, begun in 1820, was essentially completed by 1850. In that year, on the eve of Francis Wayland's restructuring of Brown University, chemistry occupied ten per cent of the curriculum, as much as Greek.[88] The decade 1850–1860 witnessed no increase in the time spent in scientific studies, only in the use of better textbooks—physics texts with increased electro-magnetism, biology texts without design and vitalism.

We pause to recognize that, at the time of the establishment of separate science courses, and separate scientific schools, the colleges had already had two decades' experience in supporting the costly scientific disciplines. They employed three or four times as many science professors as they had twenty years previously. Yet about Latin and Greek there was no question but that the colleges were going to retain what still remained of their study for many decades. Neither was there any question about the usefulness, necessity, or encouragement of science; but now that the practical applications of science had become so diffuse, the college was forced to reassess its dedication to varied mental disciplines with a uniformly prescribed course and to consider how science alone might be supplied. As all knew:

> The knowledge of chemistry that can be acquired from the course of lectures given in any of our Colleges . . . is sufficient to open to the general student one great department of the works of God, to give him the principles, and enable him to

[87] MS, George W. Benedict to the Corporation of the University of Vermont, Aug. 4, 1847, University of Vermont Archives. The Corporation called in three eminent scientists to appraise Benedict's remarks and made appropriate adjustments in staff and supplies.

[88] Brown University, *Catalogue* (1850–1851), p.15, calculates 479 hours in the college course, with 49 in Greek and 48 in chemistry.

> bring them into harmony with the rest . . . but it may be, and is superficial, and inadequate to the wants of the practical chemist.[89]

The college course, already crowded, could not absorb agricultural chemistry or the other scientific subjects that the advancing industrial age suggested. Therefore, alternatives had to be seriously considered.

[89] Mark Hopkins, *An Address delivered in Boston, May 26, 1852, before the Society for the Promotion of Collegiate and Theological Education at the West* (Boston, 1852), p.9.

VII. Utility and Science Courses: Could the College Be Something More?

THE college of the 1840's inherited both the classical curriculum that had been so altered over the previous decades and the nagging suspicion that the additions and corrections in science courses had failed to satisfy the popular mind. However great had been the progress of the institutions or however unjustified some of the public complaints, the production of skilled technological workers, which many had grown to equate with the purpose of educating their country's youth in science, was still not part of the college's program.

Plotting any new course for the college was made difficult by inherent working-class abhorrence for intellectual expertise, no matter how practical the end for which it might be directed; and the difficulty was compounded both by the financial problem of balancing any expansion of faculty and purpose with the reality of tuition revenues, and by the purely academic matter of relating any new scheme to the historic pattern of the liberal arts program itself. Remarkably, many colleges were willing to let such problems work themselves out in practice. In the late 1840's and 1850's, when deliberations on expansions to the liberal arts colleges were at their most intense, colleges undertook many since forgotten schemes which promised to extend the audience for a college education but which failed in the attempt.

Intellectual utility was a favorite nineteenth-century expression for the eventual use to which mental training was applied. But while academicians could employ this idea of intellectual utility to science as well as to other disciplines, the public rejected "mere intellectual utility" and insisted upon that more easily understood material utility to which few doubted that science was inextricably linked. According to one college president, the most significant change in the first half of the nineteenth century had been the "idea of the subjugation of the powers of nature to the use of man."[1] While progress in science could have different meaning for the gold speculator and the college geologist, it nevertheless expressed the hope of many for a better world and "has now become part of the faith of men every where who are influencing public opinion and thought."[2] The welter of party politics, sectional disputes, and variant social and economic theories could not drown out the praises of scientific progress heard everywhere; indeed, Nicholas Biddle, even in the midst of the Bank War, paused to celebrate the diffusion of scientific faith throughout the community.[3] "Science has now passed from speculation into life," as one student asserted; "it is sought as a mighty source by which nature is not only to be opened to thought but is to be subjected to our needs."[4]

Only a small intellectual minority sought to condemn this connection between science and the practical material arts it was thought to sustain. Merely a few poets—and not priests as is mistakenly suspected—supported Thoreau's belief that nature was designed for man's wonderment and not his use. The technological results of science, they argued, diverted man's attention from his moral obligations, not so much to God as to his fellow man and to himself. How was man to contemplate the good, the true, and the beautiful while engaged, for instance, in building locomotives? Professor Henry Reed of the University of Pennsylvania, who introduced the poetry of Wordsworth to America, complained that philosophy was "becoming ultra-utilitarian," that "the improvements in machinery . . . stimulate the passion for accu-

---

[1] Mark Hopkins in introduction to Emerson Davis, *The Half Century* (Boston, 1851), p.xvii.

[2] Edwin Harwood, *The New Civilization and the New Speculative Thinking: An Address before the Society of the Alumni of the University of Pennsylvania . . . Dec. 22, 1859* (Philadelphia, 1860), p.6.

For a recent article on the relationship between science and life, see Edward Lurie, "Science in American Thought," *Jour. World History* 8 (1965): pp.638-669.

[3] Nicholas Biddle, *An Address delivered before the Alumni Association of Nassau-Hall, on the Day of the Annual Commencement of the College, September 30, 1835* (Princeton, 1835), p.22.

[4] MS, John Price Durbin, Jr., "Liberal Tendencies of the Age," A commencement essay, Aug. 3, 1853, Wesleyan University Archives.

mulating wealth to an intensity that blinds its victim to the havoc he may make in the happiness of his fellow beings."[5] Ralph Waldo Emerson warned that increase in wealth, population, and scientific improvements in the arts ultimately operated to "impoverish the farm, the school, the church, the house, and the very body and feature of man."[6] And a supporter of libraries over laboratories for the Smithsonian bequest brought the argument against science before Congress:

> Sir, a laboratory is a charnel house, chemical composition begins with death and experiments are but the dry bones of science. It is the thoughtful meditation alone of minds trained and disciplined in far other halls, that can clothe these with flesh and blood, and sinews, and breathe into them the breath of life.[7]

But however elegant the minority who contended that the age was becoming "ultra-utilitarian" and abandoning more legitimate activities of the mind, the majority shared a different view in demanding an "American system in the arts and sciences,"[8] in seeking to make formal education more responsive to the popular association of nature's laws with man's machinery. Nationalism and even the Protestant work ethos could be called upon to make the denial of scientific expansion seem heretical. As science and American democracy became almost equivalent terms, a word raised against science sounded as a plea for aristocracy, a deliberate attempt to thwart the social ambitions of the lower classes. Denison Olmsted echoed this sentiment when he insisted that

> science, in very nature, tends to promote political equality; to elevate the masses; to break down the spirit of aristocracy; and to abolish all those artificial distinctions in society which

---

[5] MS, Henry Reed, "A lecture on the Literary Opportunities of Men of Business, delivered before the Athenian Institute and Mercantile Library of Philadelphia, April 3, 1838," Rare Book Collection, University of Pennsylvania Library.

[6] Ralph Waldo Emerson, *The Method of Nature* (Boston, 1841), p.13.

[7] *Speech of Mr. Marsh of Vermont, on the Bill for Establishing the Smithsonian Institution, delivered in the House of Representatives of the U. States, April 22, 1846* (Washington, 1846), p.14.

[8] Jacob Collamer, *Oration delivered before the Phi Sigma Nu Society, of the University of Vermont, Burlington, August 6, 1828* (Royalton, Vt., 1828), p.11.

> depend on differences of dress, equipage, style of living, and manners; to raise the industrial classes to a level with the professional; and to bring the country in social rank and respectability, to a level with the city.[9]

The popular conception of science, however, not only held undoubted faith in the ability of science to glean the objects of nature for economic benefit, but, indeed, probably believed that this was science's only purpose. Such a simple philosophy of investigation, maintaining that one could always discover something useful by looking at the right objects, was often related to the popular scientific utopia postulated two centuries earlier by Francis Bacon.[10] The appeal of Bacon's method of induction lay not only in his simple suggestions of how and why scientists actually "gathered facts" but also in his identification with other American social values. Bacon, along with Newton and others, fit into the Protestant Anglo-Saxon tradition which nativist orators so frequently extolled in an age of increasing Catholic hostility. When, for instance, Francis Wayland listed those "Christian philosophers" who had avoided "infidelity" in advancing knowledge, he, of course, included Bacon but omitted Descartes.[11] And northern divines found further appeal in Bacon's belief in the doctrine of the unity of nature.[12]

The academic community's more scholarly pronouncements on the method of induction, however, related a different story. One professor told his classes that

---

[9] Denison Olmsted, "On the Democratic Tendencies of Science," *Amer. Jour. Education* 1 (1856): p.165.

[10] I am not concerned here with Bacon's own inconsistencies or with the history of his philosophy since, principally, it was only the name of Bacon that so appealed to the American public. I believe that George H. Daniels in *American Science in the Age of Jackson* (New York, 1968), pp.63–67, *passim,* has flawed an otherwise fine study of national science by insisting that scientists were really as Baconian in their practice as orator Edward Everett (p.63) claimed.

[11] Francis Wayland, *A Discourse on the Philosophy of Analogy* (Boston, 1831), p.17.

For further nativist remarks on the exclusive Anglo-Saxon character of our American philosophical heritage, see Edward Hitchcock, *The American Academic System Defended* (Amherst, 1845), p.9 and Simeon North, *Anglo-Saxon Literature* (Utica, 1847), p.30.

[12] For instance, see William Adams, *Christianity the End and Unity of All Sciences and Pursuits* (New York, 1847), p.13.

> Bacon's rules are however not worth much, and the greatest philosophers disregard them &c. Induction however can be learned by almost any one. A Philosopher [scientist] must have some imagination.[13]

Teacher-scientists revealed in their lectures an awareness that scientists did not operate in the fashion that mistaken ideas about Baconianism would lead one to believe.[14] Even the theologian Horace Bushnell recognized that "the inductive method, sometimes called the Baconian, is commonly represented in a manner that would make the philosopher the dullest of beings, and philosophy the dullest of all drudgeries."[15] Similarly, the common belief that good science proceeded on the Newtonian dictum that "hypotheses" should be avoided was denied in the classroom by scientists who knew that "It is difficult & not desirable to avoid forming such opinion."[16]

But the science teacher himself could do little to clarify popular images of the nature of his work or to correct, more specifically, the naive belief that fact-gathering leads to technological improvement. In a limited sense most scientists, like the public that misunderstood them, also believed in the derivation of some ultimate if not immediate practical uses from scientific labor. Few could have failed to notice the changes in large-scale industrial processes that their tinkering had already inspired. And although in the classroom "no teacher would be tolerated, who should more than inciden-

---

[13] MS, Henry C. Cameron, "Joseph Henry Lectures on Natural Philosophy [1846–1847]," Princeton University Library.

John Foster informed his classes that Henry had told him induction consisted of "guessing and then verifying by experiment": MS, Wm. P. Chanbers, "John Foster's Lectures on Natural Philosophy to the Junior & Senior Class of Union College, 3rd term, 1847," Union College Archives.

[14] For an identical opinion by a student on the same point, see J. E. Ford, "The Nebular Theory," *Williams Monthly Miscellany* 1 (1844): p.9 insists that the "bounds of science can only be enlarged" by rejecting Bacon and calling on imagination.

A more sophisticated denunciation of Baconian empiricism appears in George P. Marsh *Human Knowledge: A Discourse delivered before the Massachusetts Alpha of the Phi Beta Kappa Society, at Cambridge, August 26, 1847* (Boston, 1847), p.27, *passim*.

[15] Horace Bushnell, *Work and Play* (Cambridge, 1848), p.31.

[16] MS, Edward Hitchcock, "Introductory Lecture on Chemistry Delivered in Amherst College [1826–1843]," President Hitchcock Collection, Amherst College Library.

tally allude to any common use [of his investigations],"[17] there was no political necessity to flaunt such sentiment in the face of prospective supporters who expected American scientists to be practical minded and unconcerned with research divorced from application.

Thus, while teaching was usually confined to principles and controversies, project proposals and other formal justifications of science and scientific education capitalized on the popular conception of scientific utility. In listing the "advantages which will result from the [New York] Geological survey," three leading geologists chose the following order of presentation"

> First, *utility* as regards our most common wants
> 2^{d}, the advancement of sciences
> 3^{d} the moral or mental influences[18]

While colleges themselves applied some of the same rationalizations to science education, "*utility*" (underscored) had certainly never been at the top of the list. Yet this very utility was clearly such an ever-present idea, with such widespread confirmation, that the college could not fail to feel its influence. "It would seem strange if the revolution in our ideas [about science] had left the department of education untouched, and if the strong tendency of the American people for the practical, had not modified very considerably, older views upon the proper course of mental culture for the young."[19] Specifically, it seemed that "Physical science, the splendid instrument of modern progress, must become, if it be not already, the controlling study of the age."[20]

Thus, the old intellectual justifications and accommodations for science gave way to expressed desires for the practical

---

[17] Charles U. Shepard, *An Address . . . before the Agricultural Societies of Hampshire, Franklin, and Hampden Countries at their Anniversary Fairs* (Northampton, Mass., 1847), pp.21-22.

[18] MS, Lardner Vanuxem, Ebenezer Emmons, James Hall to William H. Seward, [ ] 1842, Seward Papers, University of Rochester Library.

[19] George Mifflin Wharton, *An Address delivered at the University of Pennsylvania . . . before the Society of the Alumni, Nov. 23rd, 1857* (Philadelphia, 1857), p.6.

[20] Samuel B. Ruggles, *The Duty of Columbia College to the Community and its Right to Exclude Unitarians from Its Professorship of Physical Science* (New York, 1854), p.21.

results of scientific activity; and such expressions themselves occasioned a complex curricular problem for the college, which historians have often erroneously interpreted as lack of concern for science altogether. It must not be forgotten that by the 1840's science was already a prominent feature in the college, although neither it nor any other study controlled the whole. Many courses had been added in the previous twenty years, "we may almost say been *crowded* into the Collegiate course, until a long portion of time is devoted to subjects, not formerly taught."[21] By 1850 when Francis Wayland of Brown complained that this method of adding courses to the classical curriculum had exhausted its possibilities and necessitated the establishment of an elective system, his school had four scientists on a seven-man faculty and a curriculum which devoted to mathematics and science 160 out of 479 hours of instruction.[22]

Little wonder that, whether one subscribed to Wayland's radical proposals for free electives or more narrowly fixed his preference upon the standard course, all recognized that adding more scientific subjects, regardless of their theoretical or practical nature, was unattractive if not impossible. To have separated chemistry from natural philosophy, to have enlarged the number of hours devoted to astronomy, and to have added geology had been acceptable, but the process could not proceed forever. This realization which captured the classicist and scientist alike gained strength from the equally loud complaints voiced in other quarters that colleges produced neither specialists in chemistry nor experts in Latin.[23] Changing the college course principally by expanding it was not likely to bring student accomplishments more in line with these expectations than it had done in the past. Realizing this fact, the *North American Review,* whose initial response to the demands for agricultural education in 1851 was to add its "*elements* . . . into our colleges," in the same manner that

[21] *Report of a Committee of the Overseers of Harvard University, concerning the Requirements for Admission to the University* (Salem, Mass., 1845), p.6.

[22] *A Catalogue of the Officers and Students of Brown University, 1850-51* (Providence, 1850), p.15.

[23] See, for instance, Francis Wayland, *Report to the Corporation of Brown University on the Changes in the System of Collegiate Education, Read, March 28, 1850* (Providence, 1850), pp.17-18.

chemistry, geology, and biology had been accommodated,[24] four years later abandoned this hope and called for a complete return to those "permanent studies which have been regarded by wise men, for a thousand years, as the best possible discipline for young minds."[25]

College students also suggested that the classical course was already too crowded. Other modern subjects (history, social science, modern language, etc.), which themselves had dramatically usurped a significant part of the college curriculum, joined with science to present a bewildering array of mental faculties to be trained in this busy age. As a Dartmouth student described his college in 1842:

> I have to work hard. We have from five to six lectures a day, each one hour long. I am as well pleased in attending Lectures as I could reasonably expect. It is difficult to retain even the smallest portion of what I hear. The Lectures commence at 8 o'clock in the morning and continue till 12 noon with an intermission of one or two minutes between the Lectures. We should think it hard to go to church and sit for four hours at a time. The fifth Lecture comes at 4 o'clock P.M. and Tuesday we have two in the afternoon, commencing at 3 o'clock making 6 in one day. I take notes and copy them off after the Lectures. So nearly my whole time is occupied in hearing Lectures and copying notes.[26]

Less diligent scholars used "to go to sleep," "read," or "withdraw themselves from the [scientific lecture] rooms by very undignified and irregular methods," a student at Yale later revealed.[27]

The elimination of this hectic superficiality growing in the college for years was perhaps another problem that establishing scientific alternatives might solve. And educational reformers were quick to suggest procedures that might be employed. The range of these extra-collegiate proposals is indeed complex but we may simplify it by distinguishing two

---

[24] "Wayland on College Education in America," *North Amer. Rev.* **72** (Jan., 1851): p.74.

[25] "European and American Universities," *North Amer. Rev.* **80** (Jan., 1855): p.151.

[26] MS, Leonard Spalding to James Farrar, Jr., Aug. 15, 1842, Dartmouth College Archives.

[27] Charles A. Bristed, *Five Years in an English University* (New York, 1852) **1**: p.96.

categories. One suggestion for special scientific education was "partial" courses or, more exactly, a partial curriculum in which students not seeking a degree for completion of the entire course were permitted to attend those same courses that were required of the arts degree candidates. The other broad category was the parallel course, which implied that a separate sequence of courses was available to those not wishing to pursue the standard course. There are, however, some subtle important distinctions among parallel courses, as they could be composed of either the regular classical courses or truly separate offerings. What distinguished a parallel from a partial course was an orientation towards a degree, the bachelor of science or some variant. In ideology, the parallel curriculum was the precursor of the later elective systems, while scientific shools began as hybrids of a parallel course and a specific kind of partial course.

Even prior to the 1840's, additions and improvements of science courses in the "classical" or core curriculum certainly had evoked other similar suggestions. While there was variation in individual colleges' experiences, common to most of them had been an early partial course, a curriculum relying almost exclusively upon the same courses as in the full arts course and leading to no degree, although certificates of proficiency in selected subjects were sometimes awarded. This opportunity to study individual courses without gaining a degree had been tried very early in the century at Princeton and Union and, beginning in the late 1820's, at such varied schools as Harvard, Amherst, Vermont, Columbia, and the University of Virginia. By 1837 when generally conservative Bowdoin College allowed its lectures to be "attended by persons not members of the College," the plan had been tried almost everywhere.[28] Only at Union, which salvaged its partial program in 1828 when it instituted a modified parallel course leading to a bachelor of science degree, was the system retained for more than a few disappointing years.

The courses which had been expected to attract partial students sometimes consisted of special technology oriented courses but were generally the same as those the full-time students took—whether French, mathematics, or chemistry.

[28] MS, Overseers Records, Sept. 5, 1837, Bowdoin College Archives.

partial curriculum

While colleges thought they were paying greater attention to the utilitarian aspects of the study of science, professionals and laymen alike were aware that these expectations would never be realized with courses that were basically theoretical and often dependent upon the mathematical sophistication that the scientific dillitante did not possess. Thus, in 1842, the *North American Review* was able to report that "No one came of what was freely offered [;] the system dragged on for a few years and then perished from mere inanition."[29]

At midcentury when colleges were agin considering these partial courses they were constantly reminded of the experiences of twenty years earlier. Harvard and others recalled explicitly that the students in the partial course had "stood in a rank inferior to that of their associates, who were pursuing the full course. In fact, the larger portion of those, who took these partial studies, were such as had failed to gain admission into the Classical departments at the examination for entering college."[30] The University of Rochester, which did not open until 1854, repeated this same experiment and soon confirmed the observations that had been made in the earlier trials of the 1820's and 30's.

> Significantly enough the main difficulty is to secure on the part of those [partial] students the study requisite to bring them to a position in science . . . equal to that taken by those who take the entire classical course.[31]

From the student view the great advantages of merely attending college without the prestige of a degree may not have been apparent. Certainly there was no plethora of occupations upon whose ladders one could necessarily climb with a course or two in college science. If one were going to college, it made more sense to stay long enough to be labeled "liberally" educated, a factor that would later plague the land-grant schools in their early years. If colleges could raise the poor to a higher social station, then certainly a partial course was no route. Beneficiary funds or scholarships, which by this

---

[29] "College Education," *North Amer. Rev.* 4 (1842): pp.302-343.

[30] MS, Jared Sparks to Jeremiah Chamberlain, May 10, 1851, College Letters, Harvard University Archives.

[31] MS, Martin Brewer Anderson, "Report to Trustees," July 10, 1860, Anderson Papers, University of Rochester Library.

time aided many poorer youths through college, were almost never available to anyone not fully matriculated. Partial courses had been an enticing frill; but unconnected with a degree at one end of the academic spectrum or with entrance requirements at the other, they had alienated both the prospective audience and the professors.

Thus, when the ever-growing complexity of scientific, technical, and literary learning forced the college of midcentury into another round of evaluation, two previous experiences were already understood. First, merely adding new courses for everyone, already a quite unmanageable process, even if it had been a successful temporary solution, was not a likely procedure for further reform. Second, partial programs built up from exactly the same courses as given in the college proper, would be attended by ill-prepared students, if by any at all, and thus also aroused skepticism. Integrating special courses into the collegiate structure could only be accomplished if electives or extensive freedom of choice in the college actually led to a degree. With its options for upperclass students to choose either a bachelor of arts or bachelor of science program, Union College had provided a model since 1828, but only a very limited one at that, since its two programs missed being identical only by the replacement of advanced classical language studies with courses in natural history and advanced mathematics. The scientific course, fully prescribed by statute, was, in effect, a regular college course containing a little more science than the usual. Since the University of Virginia was also known to permit choice, others looked to her to see whether more appealing results were apparent.

Therefore, when Frederick Barnard, professor of mathematics and astronomy at the University of Mississippi (later president of Columbia) was trying to decide how to include such topics as analytical chemistry in the college course, he wrote to the president of Harvard for his opinion of the college at Virginia, not for an explanation of Harvard's scientific school.[32] Could one actually maintain a learning environment where the student chose what he wanted to study, studied it as long as he pleased, and still received the

[32] MS, Barnard to [J.] Walker, Aug. 5, 1854, College Papers, Harvard University Archives.

abandoned for electives

impress of a college degree? Harvard, which was optimistic enough to have already established a scientific school outside the college for these new topics, sent Barnard the disheartening reply "that in American *Colleges* [my italics], neither the age, nor the number of the teachers are such as to make the unrestrained Elective System either advisable or practicable."[33] The faculty suggested that Virginia added nothing to the satisfaction of vague popular expectations and that one should therefore be satisfied with "Scientific and Professional Schools connected with the College proper."[34]

The explanation for such a response, of course, lay in the nature of the Virginia experience itself. The University of Virginia had a number of "Schools" and the student studied in as many of these as he wanted, a system in which professional and undergraduate education were effectively merged, since the candidate for an arts degree took the identical chemistry course as the candidate for the medical degree. However, Virginia's Schools were nothing more than one-man departments, an instructional staff no larger and in many cases even smaller than at northern schools. Since a degree was conferred only upon students who had studied in substantially the same "schools" (really courses) as he would have attended in any other college, the requirements for a degree were actually no more flexible than elsewhere. In fact, they were even more strict in that the candidate for the master of arts (at this time the only non-professional degree granted) had more courses to take than his counterpart in the North. Virginia, strikingly enough, complained that it was in the North that too much freedom was allowed and argued that by granting the master's degree for mere residence, the other schools "depart [ed] from the stern requisitions of our university."[35] As for the hope that others had expressed in expecting Virginia's plan to attract great numbers of youth seeking the practical benefits of science, there was only failure. Very few students studied science at Virginia who were not also candidates for a degree, usually the medical degree that elsewhere was granted in a separate medical school.

---

[33] MS, Walker to Barnard, August 18, 1854, College Papers, Harvard University Archives.

[34] *Ibid.*

[35] "Report from the Committee of Schools and Colleges . . ." [1845], printed copy, University of Virginia Archives.

As a result, Barnard, himself a scientist, came to a sharp decision about the role of the college. He asked it to abandon such newer demands as chemical analyses and engineering altogether and to place them in special schools. And he went even further in suggesting that "until higher Universities rise up among us, let [even] the various branches of Natural History find their encouragement among associations of men whose taste lead them to their cultivation."[36] Thus, most came to despair of any solution that did not provide new programs positively distinct from the college.

Educational thought was to remain devotedly fixed upon the idea of the central liberal arts program during the entire range of extra-liberal proposals to meet professional and vocational ambitions. When President Wayland of Brown, the most outspoken critic of the course accretions, seemed to be saying in 1850 that a college ought to be a marketplace and sell whatever the student wants to buy on whatever terms he wants to buy it, the aging pioneer science educator Chester Dewey, now involved in planning the University of Rochester, asked what would happen to the "required prescribed collegiate course."[37] And President Sparks of Harvard reminded Wayland that educational experiments were fine only so long as they also "preserve the integrity and thoroughness of your old Academic course for those who may wish to pursue it."[38]

The foundations and operations of the programs established to relieve the pressures from the scientific sections of the academic course form a fascinating story of transitional development from the college to the more varied forms of the university. No historical synthesis has yet captured the texture of complexity of this movement, and this study does not pretend to fill the void. Clearly such developments were of importance in training many later scientists and others destined for technological occupations. But within the framework of the college described here, a scientific course or school movement marked the end of the era in which the range of

[36]"On Improvements Practical in American Colleges," *Amer. Jour. Education* 1 (1856): p.270.

[37]MS, Chester Dewey to Rev. Dr. Wayland, June 14, 1850, Brown University Archives.

[38]MS, Jared Sparks to Francis Wayland, Oct. 11, 1851, College Letters, Harvard University Archives.

required undergraduate studies had been constantly expanding, as well the beginning of limited electives or graduate education. Most significantly, it also marked the end of serious attempts at utilizing the college courses for partial students.

Understanding education outside the college proper is easier if we realize that at midcentury we are dealing with a new kind of partial curriculum, courses not given to the dominant undergraduate. In the opinion of educators, theoretical science had been proper mental training associated with no specific occupation, while vocational training was one of the worldly pursuits that had no place anywhere in higher education. But by the 1840's such distinctions were less firm and, according to a group of trustees, "we are relieved from dealing with either of these extremes."[39] There was much doubt expressed about the venerated opinion that God created "two kinds of knowledge," one for "self-culture" and another for "practical advantage" alone.[40] Thus, all colleges, large and small, those which were never to become universities and those which were, set out to deal with the fuzzy horizon between education in scientific theory and practice. The exploration, however, was only to be attempted in new programs.

The new courses were conceived out of a common desire to "qualify students for commercial, manufacturing, mechanical & agricultural pursuits," although nothing so comprehensive was ever achieved.[41] The flurry of attempt was, nevertheless, interesting. In 1848, armed with the general feeling that the college course should remain intact while different courses were added for those interested, President Edward Hitchcock of Amherst secretly suggested to President Mark Hopkins of Williams that the two schools create joint professorships of agriculture and other technical subjects. But Hopkins and his Williams, the apotheosis of the rural nineteenth-century

[39] *Report of a Committee of the Trustees of Columbia College Appointed to Consider and report the Subject of the Removal of the College, A Change in the Collegiate Course, the Establishment of A University System, &c.* (New York, 1854), p.13.

[40] Francis Wayland, "The Education Demanded by the People of the United States," in the *Celebration of the Fiftieth Anniversary of Dr. Nott's Presidency of Union College, July 25, 1854* (Schenectady, 1854), p.26.

[41] MS, Minutes of the Trustees of Dartmouth College, July 23, 1844.

college, actually wanted to endow their own agricultural department complete with an operating farm.[42] Princeton tried to find a professor to teach special applied courses in 1853.[43] Wesleyan, not content with just civil engineering, ran a normal school for teachers from 1840 to 1843. Union College wanted to hire a professor of agriculture among others; but, unable to fill any of the positions, it settled for a separate analytical chemistry laboratory.[44] The University of Virginia finally provided the separate course and practical training in chemistry that many mistakenly thought it had been offering earlier,[45] while such geographically diffuse colleges as Amherst, Dickinson, and Brown also supplied separate laboratory instruction to individuals not connected with the college course. The machinations to give structure to the better known Ivy League schools of science were going on at the same time.[46]

Thus shadowy plans for scientific education between the obviously theoretical and the obviously practical (which were to remain outside the classical course) were finding widespread expression in the old college and giving rise to new problems of administration, curriculum, and finance. A Harvard Scientific School where one could study the "economics of brick masonry" and "materials for laying a mile of track for a railway," as well as some of the most advanced scientific courses in the country, strikingly demonstrated that the role of the new programs was not yet adequately defined.[47] William Barton Rogers, professor of

[42] MSS, Edward Hitchcock, Confidential Letter to Mark Hopkins, July 12, 1848; Mark Hopkins to Hon. Jos. White, Feb. 3, 1853, both in Williams College Library.

[43] MS, Minutes of the Trustees of Princeton College, Dec. 20, 1853.

[44] MSS, Minutes of the Trustees of Union College, July 25, 1849, and July 21, 1855.

[45] MS, Minutes of the Rector and Visitors, March 12, 1858, University of Virginia Archives

[46] It is not my intention to trace the history of each of the scientific schools. There is a two-volume history of Yale's school alone: Russell H. Chittenden, *History of the Sheffield Scientific School of Yale University, 1846-1922* (New Haven, 1928). The dates on which new degrees were first offered, etc., are easily found in any college history. It would be a significant and necessary contribution for an historian to deal in a critical fashion with all the scientific schools. I, however, am interested here only in a few of the problems they raised for the college course. My thesis is that these schools would never have been possible if there had not already occurred a significant intellectual accommodation to science.

[47] MS, Charles A. Packard, "Formulae and Facts—Lawrence Scientific Schools. Harvard University . . ." June, 1848, Bowdoin College Archives.

natural philosophy at the University of Virginia (1835–1853) and later the first president of the Massachusetts Institute of Technology, at least understood this situation when he predicted that the college's special courses and scientific schools would "never command *numbers*" of students interested in the theoretical study of the practical arts, whatever else they might become. Rogers believed that existing colleges would continue to teach basic science courses, with perhaps only a bit more emphasis on practical application, even when they created special science courses or even special schools. He preferred completely new schools for the systematic study of engineering topics or a science of engineering, in which all the remaining pretense at mere intellectual justification for special courses was removed.[48] But the president of Harvard could insist, at least, "that the old Collegiate department is preserved and conducted on its original principles," namely to provide only the basis of professional education, not professional education itself, neither in law, medicine, theology, nor science.[49] Henry Tappan, president of the University of Michigan, whose reforming voice at this time as much saved the college as helped to establish the university, offered this analysis of the incipient attempts of the colleges to present alternatives:

> This general movement of the college towards a higher position, by adding more studies to their curriculum, by endeavoring to shape themselves to more numerous classes of students, by introducing voluntary courses of study, by attempting lectures in the more advanced branches of study, and by assuming the name of University, is not a mere freak of ambitious folly, but an attempt to meet the demands of the age . . . indications of an all pervading influence which is striving in various ways to become realized. Now everything appears crude and disjointed, and sometimes even grotesque; the fused elements are running in every direction until they find the molds which are to give them proportion and symmetry.[50]

---

[48] MS, William Barton Rogers to Henry Darwin Rogers, March 29, 1848, University of Virginia Archives.

[49] MS, Jared Sparks to Charles Caldwell, Nov. 20, 1851, College Letters, Harvard University Archives.

[50] Henry Tappan, *University Education* (New York, 1851), p. 80. Here, as throughout, my concern has been with the impact of science upon the college and not with the entire range of educational reform. Richard J. Storr, *The Beginnings of Graduate Education in America* (Chicago, 1953) is still the best source for ante-bellum efforts at graduate education.

At larger city institutions like Harvard, Yale, Columbia, and Pennsylvania, plans for education other than liberal arts shortly found some "proportion and symmetry" in permanent engineering and graduate schools. At the smaller, primarily rural schools, most of these programs fell into neglect after the War when the land grant college endowments paid for them elsewhere.

The eventual success of only a few extra-collegiate scientific programs raises interesting questions about the availability of faculty and funds when intention had already been demonstrated. The special scientific program undertaken with great enthusiasm in 1851 at Brown wobbled in 1852 when William Norton left for Yale, and died by 1855 when no replacement could be found. Union College was so desperate for an analytical chemist that it sent Charles Joy to Europe for study with a salary higher than that of the other professors, with the final result that the position could not be filled when Joy transferred to Columbia in 1857. The names of a very few agricultural chemists and engineers turn up repeatedly in the records of the many institutions who wanted to employ them. Harvard, the richest, was also aware that the "active services in the field of most of those eminently qualified as engineers" made it generally "impossible" to fill the professorships that the schools were willing to establish.[51] We may even suggest that, just as many well-established colleges failed to find the funds to compete in the expensive game of scientific expansion, so may have hundreds of ante-bellum colleges failed completely because they could not afford to sustain science even within the liberal arts course.[52]

Financial support, of course, came from donors who had made their fortunes in many ways and who doubtlessly sometimes had different intentions from those educators who would translate their trusts into practice. But however they

---

[51]*Report of the Committee of the Overseers of Harvard College, appointed to Visit the Lawrence Scientific School in 1849* (Cambridge, 1850), p. 8.

[52]I emphasize that the latter suggestion is only conjecture for which I do not have any supporting evidence. The starting place for the history of the many unsuccessful ante-bellum colleges is Donald G. Tewksbury, *The Founding of American Colleges and Universities before the Civil War* (New York, 1932).

The explicit statement that it was only the "amount of the *teaching force*" which prevented colleges from teaching "every branch of human knowledge" is found in Henry Vethake, *An Inaugural Address at Washington College, Feb. 21st, 1835* (Lexington, Virginia, 1835), p. 16.

used their support once donated, the colleges' willingness to expand preceded any certain knowledge that the funds would be found or where they would come from. Dartmouth, for example, hoped to establish separate professorships of "trade & commerce," "manufactures & mechanics," and "agriculture" before it had any idea whether their support would come from finance, factory, or farm.[53] Harvard was making bold preparation for study beyond the college in all academic areas through a school of science in the *Wissenschaft* sense, when Abbott Lawrence's funds and the dynamism of a small band of Harvard science professors succeeded in eliminating the professors of the literary and philosophical subjects that had appeared on the original roster. The story at Columbia and Pennsylvania had similar elements, where elaborate schemes for both scientific schools and graduate training were produced at midcentury with little hint of which would be funded. Shortly thereafter, the University of Rochester decided to teach agriculture, foundry work, and mining operations while still perplexed about "who would be willing to make such foundations."[54] Perhaps only the historical accidents of the new substantial private fortunes given to some schools and not to others, and the application of public funds to public institutions, prevented a larger number of existing colleges from playing larger roles in the development of technical education.

The scientific courses were expensive for the student as well as the college. In the early days of partial courses of the late 1820's, when the special students took the same courses as those offered in the college, they paid only a partial portion of the tuition bill. However, when a special professor was hired to teach analytical chemistry or engineering to only a few students, students fees had to bear the cost of these initially unendowed professorships. Fees for the courses were always higher than for the college course, and it was even possible for a special chemistry course alone to cost five times as much as the entire college curriculum. These inevitable arrangements of college cost accounting must have contributed to some of the later suspicion that colleges were the domain of the rich;

---

[53] MS, Minutes of the Trustees of Dartmouth College, July 23, 1844.

[54] MS, Martin Brewer Anderson, "Report to Trustees," July 10, 1860, Anderson Papers, University of Rochester Archives.

for while colleges were, in fact, cheap, scientific schools were not. Somewhat prophetically, it was even suggested in 1849 that adding to the college, even after large donations, was giving a deceptive appearance of wealth, because the college would forever have to wage a battle to find the funds to support these enterprises, inevitably raising the tuition of classical college students.[55]

There was little in the disposition of the ante-bellum college that would have made it unreceptive to any proposal for public utility if the public purse had been willing to pay for it. The hope for expansion in this direction was there, even if the unique American system of diffusing higher education and its resources into so many colleges prevented most institutions from producing more permanent results. But, for the meantime, many colleges did not have the money to expand beyond a liberal arts course, which, for all its faults, was still the cheapest solution known. They continued to increase the level of scientific instruction in the classical course as far as they could and demonstrated none of that ill will towards its study that was later to characterize the response of a wealthy eastern establishment in the Gilded Age.

---

[55] [S. A. Eliot], *A letter to the President of Harvard College by a Member of the Corporation* (Boston, 1849), p. 35, *passim.*

VIII. A Broader Social, Educational, and Historical Perspective

THE transformation of the curriculum of the ante-bellum college did not occur in a vacuum; it was related to the aspirations of the students and faculty, as well as to the society which the institution relied upon for support and recruitment. Through its own experimentation, the college provided fertile soil for the more sophisticated and costly structures of graduate and scientific schools that followed its own development.

The students who attended the traditional colleges just before alternatives to their strict programs became widespread were obliged to devote more time to science than they ever had before or would again. Electives, and then majors, heralded a modern age of decreased attention to scientific studies for the humanist, while allowing for the production of more professionally oriented scientists. Yet even during the period preceding such innovation, the students, "surrounded with the apparatus and breathing the atmosphere of advancing science . . . [and] familiarized to the idea of discovery . . . leave the beaten track and begin to plume their wings for original and independent flight."[1] Thus, a recent study of a group of selected scientists active in 1846 shows that "four out of five went through no special scientific or technological course aside from a regular college curriculum," and that among more than fifty per cent of the scientists, school was of primary influence in their choice of career.[2]

Of course, few of the country's youth benefited directly from

[1]Alexander C. Twining, in *Addresses delivered at the Inauguration of the Professors of Middlebury College, March 18, 1839* (Middlebury, Vt., 1839), p.44.

[2]Robert V. Bruce, "A Statistical Profile of American Scientists, 1846–1876," paper read at the Symposium on the History of Science in America, Northwestern University, March 29-April 2, 1970.

the college's increasing predisposition toward science, for less than one per cent of the college-age population attended these schools.[3] Consequently, it is often incorrectly assumed that the student body was aristocratic without a clear definition of the meaning of that term. This misconception finds support in the fact that the old private northeast colleges did achieve a relatively homogeneous social composition in the student body later in the century, the assumption being that it had always been so.[4]

Actually, in the first half of the century, "the sons of the poor always far outnumber[ed] the sons of the rich."[5] Reading the dusty records of the trustees of schools like Bowdoin, Dartmouth, Brown, or Pennsylvania, one cannot help but be struck by students' pathetic pleas for postponement of payment of the few dollars needed to cover the college costs, by the trials of the many students who earned their way through college by teaching in the district schools during winter recess, and by the administrations' concern to provide scholarships for the destitute. For example, close to forty per cent of all college students in New York state received free tuition in 1859.[6] A writer at midcentury tells us:

> In point of fact, full three-fourths of the members of the country colleges are from families with small means, if not absolutely poor. Of the whole number annually graduated in this country, a large majority are of comparatively humble parentage, many of them dependent on their own resources

---

[3]Among the nineteenth-century journals providing statistics from which such a conclusion can be drawn are *University Quarterly, Niles Register, American Almanac, American Quarterly Register,* and *Journal of the American Education Society*. My own figure of "less than 1 percent is derived for the year 1860 from data presented in F.A.P. Barnard, *Analysis of Some Statistics of Collegiate Education* (New York, 1870). For a group of northeastern states having the highest college attendance rate, Barnard concludes that, for that region alone, 1 in 71 or 1 in 100 (depending on the calculation) of "age eligible youth" attended college. For the country as a whole, it was, of course, a much smaller percentage (p.14).

[4]George E. Peterson, *The New England College in the Age of the University* (Amherst, Mass., 1964) is a provocative study dealing with the college's late nineteenth-century social adjustment to electives, Progressivism, and New England wealth.

[5]George Wyllys Benedict, *New England Educational Institutions in Relation to American Government* (Burlington, Vt., 1844), p.44.

[6]MS, Faculty Meeting Record, July 28, 1855, University of Vermont Archives; the colleges surveyed were Columbia, Union, Hamilton, Hobart, Rochester, and the public-supported University of the City of New York.

> . . . if there by any partiality in the distribution of its blessings, it is in favor of the middling and lower classes: it is emphatically the poor man's institution.[7]

Even if we include such subtle factors as the loss of income occasioned by four years of college as a deterrent to matriculation, we cannot conclude that the ranks of the college were closed to the poor or that the sons of what social aristocracy existed were the college youth of the day.[8] "If by aristocracy be meant a stupid and pretentious caste, founded in wealth, and birth, and an affection for European manners, no charge could be more preposterous," insisted President Eliot of Harvard; "the college is intensely American in affection, and entirely democratic in temper."[9] Recalling the first part of the century in 1899, President Hadley of Yale wrote:

> differences in wealth throughout the community were less conspicuous than they are today. College education was so cheap that it fell within the reach of all. Most of the students were poor.[10]

"College education in this country is cheap enough, so cheap that no one can reasonably complain of it on that score," insisted one of the most outspoken contemporary critics of the college.[11] Prior to the Gilded Age, the elitism of college

---

[7]Charles Haddock, *Collegiate Education* (Boston, 1848), pp.7–8. Edward Everett develops this same theme in his *Speech in Support of the Memorial of Harvard, Williams, and Amherst Colleges* (Cambridge, Mass., 1849), p.23. Similarly, a British historian has challenged some of the assumptions made about the social origins of students in the English universities: Nicholas Hans, *New Trends in Education in the Eighteenth Century* (London, 1951), p.41 and *passim.*

[8]A contemporary made this point even more clearly: "One thing is very certain, that did not the college exist, and exist under the same condition which they actually do exist under, none but the sons of the rich could obtain that education which is now within reach of the poorest lad in the land, *if he but earnestly longs for it, and that too without injuring in him the wholesome feelings of being independent of the charity and patronage of others*" (italics mine). G. W. Benedict, *New England Educational Institutions,* p.44.

[9]Eliot's address in David Weaver, editor, *Builders of American Universities* (Alton, Ill., 1959) 1: p.30.

[10]Hadley's address in Weaver, *op. cit.,* p.45.

[11]Francis Wayland, *Thoughts on Collegiate Education,* reprinted in Richard Hofstadter and Wilson Smith, editors, *American Higher Education: A Documentary History* (Chicago, 1961) 1: p.363. College tuition in 1831, for instance, averaged about $33 a year at northeastern institutions; see *Quarterly Register of the American Education Society* 3 (1831): pp.298–299. Denison Olmsted claimed it was cheaper to attend Yale than the academies, in "On the Democratic Tendencies of Science," *Amer. Jour. Education* 1 (1856): p.171.

education lay not in its expense but in its unpopular conception of intellectual training.

Just as college students were not recruited from the upper social classes, so a college degree did not admit one to society's greatest positions of honor and wealth. Even in the 1870's a college critic could still find that "three fourths of our national and state representatives, [and] nine-tenths of our office-holders . . . are nongraduates [of college]."[12] Consequently, neither the wealthy nor the great masses aspiring to become so thought of college in terms of financial success, most therefore still avoiding it. The exclusiveness, then, of a college education in 1850, when the rate of attendance was less than one-hundredth that of the 1970's, was not related to social class or undue attention to the needs of an aristocracy. In fact, through its insistence on a liberal arts program that made no special provision for either professional or vocational students, the college remained true to the Jacksonian theory of democracy, an essential component of which was that there be no special privilege for any group.

Thus, the American invective leveled against the antebellum college for its presumed aristocratic tendencies, both then and now, must be considered with care. It was, after all, common to argue, with the same reasoning, against the public high school movement at the end of the century, when it was considered anti-democratic to keep children in school and away from honest labor so long. At a time when the college was not yet considered a prerequisite for remunerative callings, complaints that it was undemocratic merely signified that its utility was not obvious. Science, too, whether examined within the college or whether more broadly construed, was subject to the same ignominious charge for exactly the same reason,[13] even though in reality it was beginning to open a new door of social mobility to youth of primarily rural background.

Increased opportunities to teach science in college were themselves an indication of the advantage that a college education could afford. With still no graduate schools and only a handful of popular programs beyond the liberal arts curriculum, the professors in the mid-nineteenth-century college were the major depository and dispensary of scientific

---

[12] James Morgan Hart, *German Universities* (New York, 1874), p.344.

[13] Denison Olmsted, "On the Democratic Tendencies," p.164.

expertise in education—more important even than those in the medical or engineering school. In numbers alone they had advanced from about 25 in 1800 and 60 in 1828 to over 300 in 1850.[14] Their activity and influence, while often more subtle than quantifiable, were nevertheless of great importance to the growth of an enlarged scientific culture in the United States.

The science professor, to apply the term loosely, was at best a peripheral entity in the collegiate organization of the opening decades of the century. His salary was uniformly lower than that of other professors, and his security such that he easily fitted the classic mold of the "last hired and the first fired." Consequently, the register of holders of the position in natural philosophy and mathematics reflects breaks during these early decades that are paralleled neither by other disciplines nor by science itself in the slightly later period of this study. Fortunately for the men who taught college for a year or two only then to disappear from collegiate records, their training and environment had not led them to expect either permanent or even temporary full-time employment, and they were able to fall back upon other means of support such as medical practice or lower-school teaching.

Those who had positions as professors spent the opening years of the century listening to students recite passages from textbooks whose copyright or composition dated from colonial times. Lecturing, or the presentation of explanations not contained in the textbook, was a comparatively new and little used technique. Generally unassisted by laboratory apparatus of any kind, these men performed no experiments either for the benefit of their pupils or for their own intellectual advancement.

Yet as colleges sensed the growing importance and sophistication of science, its science professors began to feel a marked increase both in status and, subsequently, in the demands that could be exacted from their schools as conditions of employ-

---

[14] These figures are based upon my estimates, starting from the total number of college professors at each period. For 1800 total, see Ellwood P. Cubberly, *Public Education in the United States* (New York, 1919), p.203; for 1828, *Quarterly Register and Journal of the American Education Society* 1 (1829): p.225; and for 1850, Abraham Suydam, "American Colleges," *The University Quarterly* 4 (1861): pp.196–202.

ment. The perennially eccentric Thomas Cooper, for instance, was only a few years before his time when he informed Transylvania College that he would teach at no institution that failed to appropriate him with an immediate $1,500 grant for equipment, supplemented by an annual $250 grant for expansion—all in addition to his salary.[15] Shortly thereafter, the best young scientists, planning to fill those positions that collegiate expansion had provided, became almost as particular as Cooper himself. Thus, in 1832, twenty-eight-year-old Charles U. Shepard, who ultimately had a distinguished research and teaching career at Amherst College, cautiously advised Benjamin Silliman, his mentor, not to suggest his name for a college position until it was known how his salary would compare with that of other professors, "how extensive is the chemical apparatus," and how large is "the annual appropriation for the expense of the chemical lectures?"[16]

While the salary inequities between professors of science and those of other disciplines were eliminated during the second quarter of the century, by the end of the period there were isolated instances of a scientist's salary being second only to that of the college president. The Union College faculty, learning that the new professor of analytical chemistry was to be given leave to study in Europe along with a salary of $1,500 while they stayed home and lived on $1,000, announced that they had never been "so moved."[17] Similarly, when the non-scientists on the Dartmouth College faculty, which was not privileged to collect extra fees from the students in the Chandler School, petitioned the trustees to eliminate the greater pecuniary advantage that the scientists now had, the trustees replied:

> we believe it just that the laborer should be paid for his service; and if, by the accident of professional qualifications, one is competent to perform those duties which another cannot, the trustees should not consider it sufficient reason to confer any

[15] MS, Thomas Cooper to Doctor Ridgely, February 12, 1816, Trustees' Records, Transylvania College Library.

[16] MS, Chas. U. Shepard to Benjamin Silliman on obverse of letter from Silliman to President [Jeremiah] Day, Sept. 25, 1832, James McDowell Letters, Duke University Library.

[17] MS, Jonathan Pearson, Diary, July 24, 1854, Union College Archives.

portion of the reward of ability and labor upon the complainers.[18]

Yet the science professor's new prestige was accompanied by new professional demands for classroom preparation and performance. By the 1820's, science journals, primarily foreign but now domestic as well, were widely distributed among the colleges in order that the active teacher might maintain closer contact with his field than the textbooks allowed. He was, in addition, expected to become "not only fully master of his special department, but also to possess knowledge in other branches of science."[19] In order to help enlarge the scanty number of textbooks available to Americans, the more widely read of these professors began preparing American editions of foreign works and, by the 1820's, were producing their own texts on the common sciences. The recitation system, suddenly deemed inappropriate to the new age, gave way, out of simple necessity and prompted by no particular foreign model, to lecture procedures much like our own. And while not every professor could boast, as did Dickinson's Spencer F. Baird, that in chemistry he gave "41 lectures and illustrated the doctrines taught by nearly 800 experiments, . . . some . . . performed by the students themselves under my direction,"[20] lectures and demonstrations became universal and student participation with the professor's assistance commonplace. If colleges did not yet expect their scientists to devote a significant portion of their time to original research, they did, at least, assume that teachers would remain constantly aware of developments in their fields, that knowledge would extend beyond that found in the textbooks, and that some "private experimentation . . . preparatory to lectures are not only desirable but necessary."[21]

Although readily accepting these new additions to their roles, science professors were not united in their ideas of how

[18] MS, Minutes of the Trustees of Dartmouth College, July 24, 1855, Dartmouth College Archives. For other examples of the increased affluence of scientists, see Edward Lurie, "An Interpretation of Science in the Nineteenth Century: A Study in History and Historiography," *Jour. World History* **8** (1965): p.689.

[19] MS, L. Agassiz to M. Hopkins, June 23, 1858, Hopkins Papers, Williams College Library.

[20] MS, S.F. Baird to Rev. Doct. How, July 11, 1848, Dickinson College Library.

[21] MS, Alexander Dallas Bache in Minutes of the Trustees of the University of Pennsylvania, Jan. 3, 1832.

much research was appropriate in a traditional college setting. Some, like Alexander Dallas Bache, looked forward to the specific encouragement of professorial research, even if necessarily in special graduate institutions. Others like John F. Frazer, Bache's successor at the University of Pennsylvania, seemed to feel that special concentration on individual research might make professors ill suited for college instruction.

Yet however effectively the midcentury reality of heavy teaching loads prevented the establishment of a subsidized research program in the college and however unclear were the forms that such a program might take, it is certain that many scientists sought recognition and influence in a wider circle than the local one of their own college. The larger scientific community was of great importance to many. As one professor admitted, "Distinction among scientific men has been with me a most powerful motive of action";[22] and another, anxiously awaiting the publication of his research results, expressed "mortification of being anticipated in part by a paper" from a different scientist."[23]

Interestingly, the common culture of ante-bellum America provided the college scientist with many opportunities for fulfilling his quest after glory and honor in a larger world than that of his own institution. Early in the century, for instance, Bowdoin's Parker Cleaveland almost single-handedly introduced the science of mineralogy into America after his own ignorance of the field had been uncovered by an inquiry from a lumbering company in Maine seeking information about the value of some rocks uncovered in the excavation of a sluice route; and years later, Cleaveland served his country again in the wholly different role as a commissioner for the survey of the North Eastern Boundary of the country.[24] James Renwick of Columbia College, one of the most sought after engineering consultants in America, was invited by speculators to search for gold in Georgia and by a president of the United States to inspect inventions intended

[22] MS, Edward Hitchcock, Private Notes, December 1843, President Hitchcock collection in Amherst College Library.

[23] MS, Joseph Henry to Benjamin Silliman, Dec. 10, 1830, Silliman Family Collection, Yale University Library.

[24] Leonard Woods, Address on the Life and Character of Parker Cleaveland, second edition (Brunswick, 1860).

to prevent steam-boiler explosions.[25] Professor Alexis Caswell of Brown University was requested by a court to apply the latest calculations of pneumatics to settle the claims of a manufacturer who insisted that a certain dam disturbed the operation of his water wheel.[26] William M. Gillespie of Union College became the country's expert on roadbuilding, while George W. Benedict of the University of Vermont served as the Northern New England authority on constructing telegraph lines. Similarly, Ebenezer Emmons managed to earn his full salary from Williams College at the same time that he collected another from the state of New York for work on its geological survey.[27] The list, in fact, may be extended indefinitely; the better known Benjamin Silliman of Yale, Alexander Dallas Bache of Pennsylvania, or Joseph Henry of Princeton were not the only professors who were engaged in the parallel developments of the college and scientific enterprise generally. In fact, every college included in this study had several professors in the period from 1820 to 1860 who were involved with the great natural history and geological surveys, the exploring expeditions, and private scientific ventures.

It is no surprise that a modern study of the most prolific American scientists of the Jacksonian period reveals that 41 out of 56 "were usually employed as professors of sciences in a college or university [*sic*]."[28] And it is, therefore, little wonder that, on the afternoon of September 20, 1848, the first formal session of the American Association for the Advancement of Science, convening in the Hall of the University of Pennsylvania, established as a category of membership "The Collegiate Professors of Natural History, Physics, Chemistry, Mathematics, and Political Economy, and the Theoretical and Applied Sciences generally."[29] Over half of the men who had taught science for five years or more at the colleges included in this study were in attendance at that meeting.[30]

---

[25] MSS, John Adams Dix to James Renwick, July 11, 1833, and Martin Van Buren to Renwick, July 19, 1838, both in Renwick Family Papers, Columbia University Library.

[26] MS, John Kingsbury to Geo. Keeley, Dec. 20, 1830, Brown University Archives.

[27] MS, Trustees Minutes, Aug. 17, 1836, Williams College Library.

[28] George H. Daniels, *American Science in the Age of Jackson* (New York, 1968), p.32.

[29] *Proceedings of the American Association for the Advancement of Science, First Meeting, Philadelphia, Sept. 1848* (Philadelphia, 1849), p.9 and pp.144-156.

[30] For short biographies of the professors considered, see the appendix.

But while changes in the college had great effect upon the recruitment of professional teachers and their students into the ranks of the national scientific community, they seem to have been less successful in gaining for the college the widespread popular support it originally had sought. Only slowly did the idea take root in the American consciousness that college education was linked to greatness and economic prosperity on the national level and to socio-economic advancement on the personal. At midcentury the "doctrine that a course of college training does not unfit a man for any business or branch of labor"[31] was, for all its hedging quality, a more enthusiastic endorsement of higher education than was common. Most Americans correctly observed "that the productive callings, so called, by which we suppose, are meant those which produce most money . . . are filled with men who have not received a college education."[32] And so long as collegiate training implied no "advantage in gaining wealth and political eminence,"[33] its attractiveness to the general population remained minimal.

To educators, the addition of science to the curriculum had seemed the most likely means of arousing greater public interest in its programs. Thus, catalogs of the 1830's advertised new studies, like the social and physical sciences, which were "more immediately connected with the business of life," meaning, of course, with the business of material gain that infected the population.[34] Certainly, every effort at reform in the 1820's, from the introduction of calculus and chemistry to the institution of partial courses within the standard curriculum, was in some way a reflection of the popular concern for utility. But in a society that had not yet equated sound science with sound national defense, in a society not yet imbued with the faith that theoretical science precedes technological advance, there was little more association of science than of Greek with the goals of a democratic republic. As one contemporary observer remarked, the populace "believeth indeed in railroads; it thinketh well of steam; and owneth that the art of bleaching by chlorine is a prodigious improvement;—but it

[31] C.A. Stoddard, "Williams' College Improved," *Williams Quart.* 1 (1853): pp.64–65.

[32] "Wayland on College Education," *North Amer. Rev.* 71 (1851): p.72.

[33] Henry Tappan, *University Education* (New York, 1851), p.64.

[34] *Catalogue of the Officers and Students of Dickinson College, Carlisle, Pa.* (Baltimore, 1836), p.19.

laughs at the profound researches into the laws of nature out of which those very inventions grew."[35]

It is only fair to the sentiments of the general public, however, to admit that the scientists or educators could easily be unconvincing in the reiteration of their belief that systematic study of nature resulted in practical social gains. The geological surveys in which they participated, for instance, were not contributions to mining technology, as the public well knew. For all their zeal, these men never paused to consider that the major innovations in the industrial technology of the first half of the nineteenth century, from textile processes to the application of steam, had all been developed by tradesmen independently of such theoretical discoveries as the scientist might call pure research. The fact that the future was to hold a closer cause and effect relationship between the progress of science and technology served in no way to lessen the effect of conventional wisdom in 1828.

Significantly, the college was unable to overcome its unpopular image simply by instituting a number of courses in applied, as opposed to theoretical, science; for the two differed less in operation than might be expected. When the University of North Carolina, for instance, was trying to decide in 1853 what kind of northern model its scientific school would employ, a professor reminded the president that the college study of applied science was pedagogically more closely related to the theoretical than the undiscriminating proponents of scientific utility would believe; in words somewhat reminiscent of William Barton Rogers' call for a science of engineering, he observed:

> The simpler application of mathematics to Civil Engineering have always been taught here. . . . The demonstration of a theorem is often short and easily understood while its application is as prolix and tedious. . . . Teaching for education [which may include practical application] and teaching for information and practice are two quite different things.[36]

---

[35] C.S. Henry, *The Importance of Exalting the Intellectual Spirit of the Nation: And Need of a Learned Class* (Burlington, New Jersey, 1836), p.37.

[36] MS, Chas. Phillips to [Pres. D.L. Swain], June 15, 1853, College Papers, University of North Carolina Archives.

Ultimately, it was such recognition that the "practical sciences" required theoretical study which formed the foundation of the scientific-technical-engineering schools of the later nineteenth century.

Little of the change in the college's program, however, would have been possible at all had it not been for the growth of the national economy in the period from 1815 to 1860. In 1815 many of the best known of our now national institutions of higher learning were local community colleges earning a barely sufficient income from student tuition and real estate sale. With no federal support at all and only little from the local and state governments, those colleges that were able to survive until 1860 did so through philanthropic donations derived from the substantial personal wealth accumulated by private citizens. In addition, alumni organizations first became popular in the 1830's when their principal task consisted of the coordination of fund-raising drives for their institutions. The thousands of dollars which furnished schools with laboratories, observatories, and other apparatus came principally from donors who had only recently attained their wealth in new industrial manufactures, many of whom—like Abbott Lawrence, who gave something to each college in his home state of Massachusetts—had not been to college themselves.

The proto-industrial economy, moreover, served the college as more than a mere source of raw capital. The diversity and multiplicity of tasks in that economy provided new employment opportunities for men trained, however slightly, in the sciences. And it provided constant encouragement to both the educated population and the colleges themselves in their ever-increasing support of scientific endeavor. In time, the reciprocal advantages that the economy and scientific activity held for one another would be greater than anyone in 1828 could have predicted.

Ironically, time would also later create the impression that the college's accommodation to the intellectual demands of science was less important than indeed it was. The educational historian himself bears responsibility for this phenomenon, for it is he who has formulated the frequently heard charge that the college nurtured a curriculum so limited in its scope and elite in its consideration as to be irrelevant for the democratic society in which it flourished. But modern scholarship has constantly undervalued the variety and extent of scientific instruction common in the ante-bellum college. Clearly my characterization of it as a home of scientific activity and an important source of diffusion of scientific culture bears little resemblance either to traditional histories of the college or to those histories of science in America that rely upon them. And the discrepancy is one which requires some explanation.

Essentially, the problem lies in the fact that, in the absence of a significant body of thorough scholarship, the real position of the ante-bellum college has suffered from wide acceptance of the undocumented platitudes of liberal academics seeking assurance that the kinds of higher education they have supported since the late nineteenth century are justified by the illiberality of their predecessors. In their praise of the scientific triumphs of graduate schools, scientific schools, and land-grant universities, the Whiggish historians have neglected to explore what preceded such specialized developments; and their assumptions much resemble those of a man who, having immersed his hand into a basin of warm water, supposes that the liquid was just heated from freezing rather than recently cooled down from boiling. While the real state of scientific instruction was more like the second than the first alternative, from the time of Frederick Jackson Turner, who wrote of it that "emphasis was laid upon the preparation for the ministry [while] the curriculum was rather rigidly required and of the old classical type,"[37] historians of American education have uncritically accepted a kind of curricular frontier thesis which insists that older collegiate institutions were by their nature incapable of responding to democratic expansionist challenges. A more recent historian remarked in that vein that "many a campus rumbled on as though nothing of importance had happened since Caesar found all Gaul divided into three parts,"[38] thereby dismissing the subtle problems of religion, ideology, education, and science which cry for recognition. Allan Nevins perhaps typifies the mud-

[37] Frederick Jackson Turner, *The United States, 1830–1850: The Nation and Its Sections* (New York, 1963), p.17; originally published in 1935.

[38] George P. Schmidt in *A Century of Higher Education: Classical Citadel to Collegiate Colossus,* edited by William Brickman and Stanley Lehrer (New York, 1962), p.55.

dling of a series of separate problems in his statement that in the 1860's a revolt against the college grew from a "rejection of the tyranny of classical and theological studies, championship of science, insistence on attention to agriculture and the mechanic arts, and—most of all—a demand for greater democracy in education."[39] The suggestion is, of course, that such terms as classical and denominational can be used not only to distinguish liberal-arts institutions from the land-grant college of the frontier, but to establish the fact that the former were anti-scientific by nature—an implication which could not be further from the truth.

To begin, the era of the "agricultural" college does not represent such a sharp break with that of the classical college as is usually implied. Surely Senator Morrill, who admitted that the term "agricultural would never have been applied to the [land-grant] institution except that it happened to suit the casual convenience of an index clerk," realized that he was using public funds only to support establishments much like the traditional colleges, which already included science.[40] Moreover, just as the term agricultural may be a misnomer as applied to the land-grant institutions, so might the terms denominational and classical be misleading in their application to the nineteenth-century liberal-arts college; for such

---

[39] Allan Nevins, *The State Universities and Democracy* (Urbana, 1962), p.2. This equation of the older colleges with anti-American values is a common theme in land grant college literature. See, for example, Frederick B. Mumford, *The Land Grant College Movement* (Columbia, Mo., 1940), pp.71-73, *passim*.

While no one has yet produced a synthesis that serves as a challenge to the traditional historiography of the American college, a number of recent authors have questioned its cherished assumptions about the place of religious culture and science in the college, the relationship of the college to the later university, and the character of the student body. See, for example, James Axtell, "The Death of the Liberal arts College," *History of Education Quarterly* **11** (1971): pp. 339-352; Hugh Hawkins, "The University Builders Observe the Colleges," *ibid.*, pp. 353-362; David B. Potts, "American Colleges in the Nineteenth Century: From Localism to Denominationalism," *ibid.*, pp. 363-380; David F. Allmendinger, Jr., "New England Students and the Revolution in Higher Education, 1800-1900," *ibid.*, pp. 381-387; Natalie A. Naylor, "The Ante-Bellum College Movement: A Reappraisal of Tewsbury's Founding of American Colleges and Universities," *Hist. Ed. Quart.* **13** (1973): pp. 261-274; Stanley M. Guralnick, "Sources of Misconception on the Role of Science in the Nineteenth-Century American College," *Isis* **65** (1974): pp. 352-366; and Jurgen Herbst, "American College History: Re-Examination Underway," *Hist. Ed. Quart.* **14** (1974): pp. 259-266.

[40] Quoted in George W. Atherton, "The Legislative Career of Justin S. Morrill," *Report of the Commissioner of Education for the Year 1899-1900* (Washington, 1901) **2:** p.1329.

institutions included no theology in their curriculum, nor did they confine their program of study, as is usually held, to the classical languages. In short, we cannot understand the nineteenth-century college, nor the part science played in its development, without a reinterpretation of the vocabulary traditionally used to describe it.

The term denominational is possibly the most misleading to be used in many a cursory account of the American college. While the majority of colleges were associated with some sect and religious motivation at their inception, even before 1800 they could not be properly characterized as proselytizers. By 1825 all the major Protestant sects had established theological seminaries or divinity school where the real work of preparing ministers took place. Some denominations required no formal training at all for the ministry; some required seminary training alone; others required a college education for entrance to the seminary. But for none did a college degree imply ordination. In fact, during the Jacksonian period, only a minority of college graduates subsequently carved out ministerial careers, and many of these were terminated by such new professional opportunities as teaching and even science itself.[41]

Little systematic theology can be found in the college curriculum; and most colleges tried to keep a steady distinction between an ecumenical denominationalism and a confining sectarianism which, it was generally agreed, "would inevitably affect the great objects aimed at in its [the college's] establishment."[42] The nineteenth-century religious ethos, then, did not permit church-oriented colleges to become educationally unsound institutions. Even the religious revivals that so often swept through the colleges as testimony to their moral worth were strictly confined to the extra-curriculum.

Although denominational colleges were founded in great numbers throughout the ante-bellum period, impetus for their establishment was more likely to be provided by the local

[41] An interesting, but severely limited study of nineteenth century college graduates is found in Bailey B. Burritt, *Professional Distribution of College and University Graduates: United States Bureau of Education Bulletin* (Washington, 1912).

[42] *Inaugural Address of William Chauvenet, L.L.D., as Chancellor of Washington University* (St. Louis, 1863), p.31.

communities than by any ecclesiastical body; and the people who initiated the criticism about these numerous schools were generally at denominational colleges themselves. Their complaint about all the small colleges of the midwest was not that they might champion the beliefs of a given sect, but that their great numbers prevented concentration into "central and competent institutions, adequately furnished and endowed, at which the languages, the arts, and sciences, may be thoroughly and extensively studied."[43]

This too was the real theme which Andrew Dixon White was to reiterate later in the century as a corollary to his famous thesis that myth and systems of theology had hindered the acceptance of scientific modes of thinking. We too readily assume that White combined his general criticism of dogmatic theology with his dislike for American denominational colleges, to conclude that religion in colleges prevented the promotion of science. As an ardent promoter of universities, however, White complained principally that the randomness of the denominational colleges hindered the development of larger, more central institutions for advanced instruction. Fully subscribing to the contemporary belief that the college should be a center of general Christian sentiment, he certainly never suggested that ecclesiastical powers either inside or outside of the American college were interested in suppressing science.[44] The first president of Johns Hopkins University, who watched this misinterpretation take form in the minds of others less careful and articulate than White, declared:

> Hostility toward scientific pursuits or toward scientific instruction has never in this country been manifested to any noteworthy extent by the religious part of the community or by theological teachers. In discussions relating to the sphere of science and religion, the teachers of religion have almost always been earnest in their approval of scientific research.[45]

---

[43] Thomas I. Wharton, *Discourse delivered before the Alumni of the University of Pennsylvania, at their First Annual Meeting on the 14th of July 1836* (Philadelphia, 1836), pp.11-12.

[44] See, for instance, "The Relations of the National and State Government to Advanced Education," *Old and New* (Oct. 1874); *Autobiography of Andrew Dixon White* (New York, 1905) **2:** p.527; and *A History of the Warfare of Science with Theology in Christendom,* edited by Bruce Mazlish (New York, 1965), p.22.

[45] Daniel C. Gilman, "Education in America, 1776-1876," *North Amer. Rev.* **122** (1876), p.224.

Science professors themselves found a religiously inspired morality imperative for the study of science and agreeable to their own thoughts and interests. The majority of them are readily identified as active members of one of the Protestant sects. Many of those who were New England Congregationalists, for example, had spent some time in their early years studying at the Andover Seminary while trying to find their bearings in a world that offered new life styles. The mere fact that such men were able to pursue their religious studies almost to the point of a career and then switch to science is, in itself, some indication that the two kinds of thought could be congenial to the same mind.

Like science, Calvinism had changed considerably since the seventeenth century, harmonizing now with all American Protestantism in no longer naively expecting that man's eyes or senses alone could reveal the workings of nature and hence make him more aware of his Maker. Leaders of theological thought now emphatically encouraged the inspection of nature with instruments, mathematics, and reflections upon the mind's received impressions. As scientists, then, Americans were empiricists, allowing both their religion and their common sense to offer strong arguments against denying the existence either of matter or that which clearly presented itself to their senses. It was the accepted credo that "empirical spirit" applied anywhere, even to religion, "but stopping short of faith, makes an age proud and critical, rather than [merely] humble and believing."[46]

Far from hinting that the study of physical science could in any way lead to infidelity, the religiously based morality that the college expressed insisted that "He [God] likes mathematical demonstrations, because they make some things certain . . . [and] philosophical investigations, because they are fitted to extend the circle of knowledge."[47] It was also commonly held that "the more fully the man of science puts himself into sympathy with . . . the Almighty . . . the more

---

[46] George B. Cheever, *Experiment as Infidelity or Faith: An Address delivered before the Society of Inquiry on Missions in Amherst College, August, 1843* (New York, 1843), p.19. In this connection, also see S. Chapin, *An Inaugural Address delivered in the City of Washington, March 11, 1829* (Washington City, 1829), p.20.

[47] Nathan Lord, *The Improvement of the Present State of Things,* (2nd ed., Hanover, N.H., 1860), p.14.

likely is he to make advances in scientific attainment."[48] Scientists bore witness to the now forgotton prophecy that scientific progress and the spread of the influence of the Christian God would always go hand in hand.[49] Edward Hitchcock, in fact, one of America's most productive scientists and a devout Calvinist, even equated his "scientific pursuits . . . [with the] means of personal sanctification."[50]

But even though science and religion expressed values in common, scientific discoveries, especially in geology, frequently fostered debates about the accuracy and meaning of the Bible among many participants; and certain features of these pre-Darwinian debates need to be emphasized. First, there are no instances in the college literature of college presidents (almost always ministers themselves) using the occasion of the debates to chastise science or to suggest any diminution in the college's commitment to its study. Second, no instances appear when science teachers withheld information from their students or expressed judgments antithetical to world-wide scientific opinion because of community demands for religious conformity. Third, all of the essential beliefs of Christianity, such as faith in man's special moral nature and the divine inspiration of the revealed word, continued to be expressed by scientists without hypocrisy or impairment of their scientific acumen. The eminent chemist Wolcott Gibbs, for instance, who was excluded from a position at Columbia College not because his scientific views were heretical but because he was a Unitarian, maintained faith in the harmony

[48] MS, Martin Brewer Anderson, "Relation of the Scholar to Practical Life," [1854], Anderson Papers, University of Rochester Library.

[49] See, for instance, MS, George W. Wendell, Jr., "Philosophical Oration: Influence of Speculative Minds. Delivered at Commencement, August 4th, 1852," Wesleyan University Archives; and Francis Wayland, *The Dependence of Science Upon Religion* (Providence, 1835), p.30, *passim.*

[50] MS, Edward Hitchcock, Private Notes, December, 1842, Hitchcock Family Papers, Amherst College Library.

On the mutual reenforcement of social values, religious faith, and the psychology of the scientist see Stanley M. Guralnick, "Geology and Religion before Darwin: The Case of Edward Hitchcock, Theologian and Geologist (1793-1864)," *Isis* **63** (1972): pp. 529-543; John C. Greene, "Science and Religion," in *The Rise of Adventism: Religion and Society in Mid-Nineteenth Century America,* edited by Edwin S. Gaustad (New York, 1974), pp. 50-69; and Charles E. Rosenberg, "Science and Social Values in Nineteenth-Century America: A Case Study in the Growth of Scientific Institutions," in *Science and Values: Patterns of Tradition and Change,* edited by Arnold Thackray and Everett Mendelsohn (New York, 1974), pp. 21-42.

of science and religion. He testified that "he believed in the entire conformity of the two revelations of scripture and nature and that he knew of nothing in any branch of science . . . which impugns the authority of the sacred writings or conflicts with their teachings."[51] This, indeed, was the essence of the mutual reinforcement of the two faiths of scientist and believer. For whether the scientist felt that the Word and the world should be reconciled or kept distinct, he found the general environment and also that of the college encouraging to his scientific work. Neither the scientific nor the religious establishments suggested that one could validly draw an "inference from . . . science to irreligion."[52] When one exuberant Professor of Greek spoke irreverently of the findings of science, a scientist on his faculty suggested that the affront would cause a "stir among theologians and naturalists [both],"[53] as indeed it did.

But if one still feels that the habitual behavior of religious enthusiasm, essentially free of any damaging intellectual content, prevented scientists and science teachers from prosecuting science according to their own professional standards, he should consider how long that Christian sentiment lasted into the age of the university, with its usually undisputed scientific excellence. As late as 1899, for instance, Arthur Twining Hadley, the first non-divine to be resident of Yale University, was examined by the Corporation and declared to be free of heretical beliefs before assuming office. The most recent and comprehensive of historians of the American university has suggested that

> inherent differences between state and private institutions in the nineteenth century can easily be exaggerated; . . . although state universities de-emphasized religion somewhat more rapidly than did private colleges, the remarkable thing is how long officially sponsored religion endured at many state endowed universities.[54]

---

[51] *State of New York. No.67 in Senate, March 10, 1855. Report of the select Committee appointed to examine into the affairs of Columbia College,* printed copy in Columbiana Collection, Columbia University Library, p.5.

[52] Mark Hopkins, *Science and Religion* (Albany, 1856), p.24.

[53] MS, Jonathan Pearson, Diary, July 15, 1855, Union College Archives.

[54] Laurence R. Veysey, *The Emergence of the American University* (Chicago, 1965), p.112.

Still, the records of the daily activities of these colleges make it clear that piety and science are not mutually exclusive. Educational history is too easily distorted by the suggestion that a society or culture which expresses religious and scientific goals cannot keep instruction in each separate and distinct. There was, after all, "no such thing as Presbyterian or Episcopalian chemistry."[55]

As the meaning of the term denominational has been misunderstood by modern historians, so too has the term classical. It is constantly placed in contradistinction to "scientific," obscuring the fact that in the nineteenth century "classical" was regularly used synonymously with "the liberal arts," thus including the scientific. The term was used to distinguish not dead languages from science, but liberal education from vocationalism and technical training—whether in law, theology, or scientific research. The usual form of the argument that we hear today—namely, that the study of Latin and Greek kept science out of the program—did not apply in America, however true it may have been at Oxford. President Mark Hopkins of Williams, who noticed this impression growing even as the classical course was altered, argued:

> if it be intended that improvements in the sciences are not ingrafted, as they are made, upon the scientific courses, or that new sciences are not introduced as the wants of the public demand; if it be intended that there is an adherence to things that are old because they are old, then however much ground there may have been for the charge formerly, and especially in England, from which this complaint is mostly imported, I do not think there is any ground for it now [1836]. It is within the memory of our older graduates that Chemistry, and Geology, and Mineralogy, and Botany, and Political Economy were either not taught at all, or scarcely at all in the college course.[56]

Hopkin's remarks were an elaboration of the same tradition expressed in the evaluative surveys of the 1820's, a tradition to which the scientists also subscribed.[57]

---

[55] MS, Charles A. Joy to Hamilton Fisk, Jan. 27, 1854, Fisk Papers, Columbia University Library.

[56] Mark Hopkins, *An Inaugural Discourse, delivered at Williams College, September 15, 1836* (Troy, N.Y., 1836), p.21.

[57] Acquiescence in the faculty psychology of mental discipline is implicit in the educational writings of many scientists. For overt insistence upon the standard liberal

The opponents of classical education were, in fact, rarely science teachers but rather the great mass of citizens. Educators were aware that the "ordinary objections to liberal education lie not so much against this or that branch of profound or elegant learning, but against all."[58] They knew that there had "been very little controversy" between literature and science, and "that the influence of Harvard, Yale, and the other older colleges . . . [had been] powerful throughout the land" in creating institutions that fostered both.[59] In short, they knew "contrary to an impression general today" that "the curriculum was not invariable in content, although rigid in its prescriptions."[60]

College education, then, most certainly involved a fusion of classical learning and science. Yet the general historian has failed to see this truth owing to superficial reading of the college records, while the historian of science in America has sinned even more by almost totally neglecting them. A belief that nineteenth-century America was peculiarly indifferent to the basic problems in science and only enamoured of its practical results has probably cemented the scientific historian's indifference to the significance of scientific culture in an educational system that only slowly produced research institutions.[61] I. Bernard Cohen, for instance, who was perhaps

arts curriculum as preparation for all callings, including that of scientist, see James D. Dana, "Science and Scientific Schools," *Amer. Jour. Education* 2 (1856): p.351; and Alexis Caswell, "Importance of the Mathematical Studies Considered as a Branch of Liberal Education," *The Biblical Repository and Quarterly Observer* **21** (Jan. 1836): pp.1-2.

[58] MS, Martin Brewer Anderson, "End and means of a liberal education," an address dated 1854, University of Rochester Archives.

[59] Daniel C. Gilman, "Education in America, 1776-1876," *op. cit.*, p.224. Compare this with the usual interpretation that science grew with stubborn opposition from classicists, as stated, for instance, by Harold Larrabee in *Naturalism and the Human Spirit,* edited by Yervant H. Krikorian (New York, 1944), p.346.

[60] Richard J. Storr, *The Beginnings of Graduate Education in America* (Chicago, 1953), p.5.

[61] The classic statement of this thesis is Richard H. Shryock, "American Indifference to Basic Science during the Nineteenth Century," *Archives Internationales d'Histoire des Sciences* **28** (1948-1949): pp.3-18. This view of American science has been admirably questioned by Nathan Reingold, "American Indifference to Basic Research: A Reappraisal," paper read at the Symposium on the History of Science in America, Northwestern University, March 29-April 2, 1970.

Obviously, I find Shryock's thesis difficult to accept, for I suspect that no society has existed which worships other than the visible products of science. A modern study of attitudes toward science in nineteenth-century England, for instance, pictures that country "as a nation personifying materialism . . . [where] the practical function was overly impressed with the scientific achievement of the eighteenth-century college, never looked beyond the first two comparatively destitute decades of the nineteenth for the yet "greater scientific movement," whose existence he even unwittingly postulated.[62] But, if in 1840 the college was not yet the graduate school of the future, neither was it that remote scholastic arena of the past.

Thus, the traditional image of the college as a citadel of rigid theological and ancient language study informed with little science detracts from the college's most fundamental character and importance. In the second quarter of the nineteenth century the college built upon its already favorable disposition toward science to create positive relationships with the growing scientific establishment. It encouraged the scientific literacy that was a prelude to greater American participation in research. It overcame some of the difficulty of an intellectual accommodation to science that permitted the modern system of scientific and technological education we so hail. A pervasive concern for the perpetuation of the scientific spirit in education grew virtually unhindered; there were few people, and fewer still in the college, who would have denied that "instruction in physical science, should hold a place, and not an inconsiderable one, in every system of education."[63] Only when that spirit attempted to satisfy the needs of the intellectual, the expert, and the utilitarian at once, would the problems arise that have plagued systems of education ever since.

---

the more understood and the more respected." Careful analysis of any other modern country is likely to reveal the same. George A. Foote, "A Study of Attitudes toward Science in Nineteenth Century England, 1800-1851" (unpublished doctoral dissertation, Cornell University, 1950), p.296.

[62] I. Bernard Cohen, *Science and American Society in the First Century of the Republic* (Columbus, 1961), p.4.

[63] "Elementary Works on Physical Sciences," *North Amer. Rev.* **72** (April 1851): p.358.

APPENDIX

I. COLLEGE LECTURES, 1847-1848

Excerpt from *Catalogue of the Officers and Students of the College of New Jersey for 1847-48* (Princeton, 1848), p.19.

In addition to the recitations of the several Classes, the following course of lectures are delivered on the principal branches of science and literature, namely, a course

On Mental and Moral Philosophy	By the President
On Greek Literature	Prof. Maclean
On Geology	Prof. Henry
On Mechanical Philosophy	Do.
On Physics	Do.
On Chemistry	Prof. Torrey
On Mineralogy	Do.
On Botany	Do.
On Astronomy	Prof. S. Alexander
On Mathematical and Physical Geography	Do.
On Rhetoric	Prof. Hope
On English Literature	Do.
On Natural Theology, including the outlines of Anatomy and Physiology	Do.
On History	Prof. Forsyth
On Architecture	Prof. Giger
On Zoology	Dr. Schanck

Gentlemen not connected with the College have the privilege of attending the above lectures by making application to the several lecturers.

II. SAMPLE OF ASTRONOMICAL QUESTIONS OF THE TYPE USED AT RECITATIONS IN THE EARLY PART OF THE NINETEENTH CENTURY IN CONNECTION WITH GLOBES

MS, Parker Cleaveland [n.d.], Cleaveland Papers, Bowdoin College Library (probably 1808-1828—the Report of the Visiting Committee, Sept. 4, 1832, MS, Bowdoin College Library, mentions a pair of worn globes that had been used for twenty years).

What place is N. Lat. 39°56' W. Long. 75°11'?
What star has Rt. Ascension 66°30' N. Declination 16°10' and in what constellation?
What is the Latitude and Longitude of Dublin, Ireland? Is the Long. E. or W.?
In what direction is London from Brunswick? & What distance in the great circle?
When it is 12'c at Brunswick, what time is it at London?
What is the sun's azimuth at rising June 21st?
What is the line of sun rising Dec. 21?[1]

[1]It does not take a scientist to recognize the child's play that passed under the name of astronomy before there were competent textbooks or astronomers. Even the last two questions, presumably the most conceptually difficult, are answered by reading a sliding scale on the terrestrial globe; thus, any semi-literate person with the globe in front of him could give the correct answers withouth any knowledge of astronomy.

III. A SAMPLE OF EXAMINATION QUESTIONS USED IN ACCOMPANIMENT WITH HERSCHEL'S ASTRONOMY, 1845

From MS, A. S. Kellogg, "Questions on Herschel's Astronomy; Given out for Examination Aug. 15th, 1845," Williams College Library. This excerpt represents about half of the whole.

To obtain the earth's diameter from the dip of the horizon.
To obtain the Latitude of any place by celestial observations.
To distinguish between diff. kinds of time: as mean, solar, sidereal, local & absolute.
Explain the construction & principal adjustment of the Transit instrument.
To convert Right Ascension & Declination into Latitude and Longitude; and vice versa.
What is the precession of the Equinoxes & its Uranographical effect.
Define parallax—and explain the method of obtaining the moon's hor. parallax. Also give the amount of this quantity.
Deduce the Sun's hor. Parallax from the transists of Venus & give the amount.
Explain Herschel's method of estimating the dist. of a fixed star; & give Bessel's estimate of the dist. of 61 Cygni.
To obtain the magnitude of the heavenly bodies.
To explain that the modern theory of celestial motions is correct.
Prove the earth's orbit to be elliptical; give its eccentricity: & the line of performing one revolution.
Define a disturbing force: & explain the manner in which these forces take their origin in the system.
State the law of forced oscillations, & give an instance of it.
Explain within what limits solar or lunar eclipses are liable to happen.
To find the line of the beginning, middle, & end of an eclipse by trigonometrical calculation.

IV. CHEMISTRY QUESTIONS, 1841 "NOTES AND QUESTIONS ON TURNER'S CHEMISTRY"

A printed list of twenty pages [n.p.n.d.] used in conjunction with the many American editions of the most extensively used textbook from 1830–1850. This copy is marked "Lewis F. Laquereme, Esq., Sept. 8th, 1841, University of Pennsylvania," Edgar Fahs Smith Memorial Collection in the History of Chemistry, University of Pennsylvania Library.

Sample questions:

How do we find the specific gravity of a liquid?
How may the correction [for unequal expansion] be made for temperature?
Wherein does affinity differ from cohesion?
Explain the separation of salts by crystallization.
What is the utility of the laws of combination?
Describe Dr. Wollaston's Scale of Chemical Equivalents.
Does the correctness of the chemical equivalents depend upon the atomic theory?
What is the composition of water?
How may this be proved?
What is the composition of the atmosphere?
What is the best mode of analyzing it?

V. BIOGRAPHIES OF SCIENCE PROFESSORS

From 1828 to 1860, fifty-nine men, appointed prior to 1853, taught mathematics or science for a period of at least five years at one of the colleges studied here. Biographies of all appear in this appendix. Even in the seven or eight cases where the men were so obscure that little biographical material remains, their lives have been included in order to eliminate bias in my presentation.

In addition to providing a reference for figures in the text who might otherwise be unknown, the biographies are intended to show how extensively teachers of science participated in the scientific activities of their world. Although it still seems strange to some that science can be carried out in so many contexts by men who are not geniuses, we must nevertheless understand that the preparation of textbooks when there were none, or the participation in geological surveys when the country was yet scientifically unexplored, were legitimate scientific endeavors in the nineteenth century. Thus, to ask why there are not more "great men" in this group of scientist-teachers is ultimately equivalent to posing Reingold's sarcastic query, "Why wasn't Sir Isaac Newton a Bulgarian?"*

About a dozen of the science professors discussed here are already well known to the historian of American science, while some are familiar even to the general historian. In such cases, a principal reference to the teacher's life is provided; but no claim is made to biographic completeness. In all cases, even in those instances where a biography seems to follow the format of the *Dictionary of American Biography,* I have been guided by manuscript material found in the libraries of these professors' colleges.

---

*Nathan Reingold, in review of *American Science in the Age of Jackson* by George H. Daniels, *Isis,* (Spring, 1968), p.120.

ABBREVIATIONS USED IN BIOGRAPHICAL APPENDIX

AAAS	American Association for the Advancement of Science
AAAS, Proc.	*American Association for the Advancement of Science, Proceedings.*
ACAB	*Appleton's Cyclopaedia of American Biography*
AJS	*American Journal of Science*
APS	American Philosophical Society
BDA	*Biographical Dictionary of America*
DAB	*Dictionary of American Biography*
NAS	National Academy of Sciences
NAS, Biog. Mem.	*National Academy of Sciences, Biographical Memoirs.*
NCAB	*National Cyclopaedia of American Biography*

Adams, Charles Baker (1814–1853). Middlebury College, 1838–1847; Amherst College, 1847–1857.

One of the enobled Massachusetts Adamses, Charles Baker started college at Yale but transferred to Amherst, where he studied under Edward Hitchcock (*q.v.*) and graduated at the head of the class of 1834. After spending two years studying theology at the Andover Seminary, he returned to Amherst as a tutor, lectured on geology at Bradford Academy, and assisted Hitchock on the geological survey of New York. In 1837 Adams accepted a teaching position at a college in Missouri but, after its failure in 1838, returned to New England to become professor of chemistry and natural history at Middlebury College. While at Middlebury, he also taught geology, gaining field experience from his work as state geologist of Vermont and eventually publishing four annual reports of this survey under the title of *Annual Reports on the Geology of the State of Vermont* (Burlington, Chauncey Goodrich, 1845, 46, 47, and 48). In addition, he collaborated with his classmate Alonzo Gray on *Elements of Geology* (New York, Harper & Bros., 1863), a textbook of limited popularity which, although completed by 1852, was not published until 1863.

The second phase of Adams's short career began when he prepared, while still at Middlebury, a catalog of local mollusca, which appeared in the *AJS* 40: pp.266–277. His interest in zoology and mollusks in particular carried him to the tropics to collect specimens. He produced about a half dozen volumes and monographs on these researches, including *Contributions to Conchology* (New York, 1849–1852) and *Catalogue of Shells Collected at Panama* (New York, privately printed, 1852), both long considered as classics in American conchology.

In 1847 Adams returned to Amherst to become professor of astronomy and zoology and to be with his venerated teacher Edward Hitchcock, after whom he named a son, but not to interfere with the geological department. He brought to Amherst his private collection of 4,500 separate species of shells with over a million duplicates; 1,500 specimens of American and European insects; 500 species of mammals, birds, reptiles, fish crustaceans, and echinoderms; 2,000 species of dried plants; and thousands of rocks and minerals acquired in the survey of Vermont. Although Amherst retained this gigantic collection, a number of select specimens were sent to the British Museum in 1850.

Extremely popular with both Middlebury and Amherst students, Adams was a member of the Association of American Geologists (later the AAAS), all principal American natural history societies, and numerous foreign societies. His career ended tragically at age thirty-nine when he contracted yellow fever during a trip to the West Indies in 1853. At the time of his death, he had already recognized the inadequacy of the old conception of "immutable species" and was groping with some of the same problems which Darwin faced.

DAB.
Seeley, Henry M. 1903. "Sketch of the Life and Work of Charles Baker Adams." *American Geologist* 32: pp.1-12.

Adams, Ebenezer (1765-1841). Dartmouth College, 1810-1833.

Ebenezer Adams, the oldest member of this group of fifty-nine, really belongs to the previous generation, although mathematically he just fits into the survey. Born in 1765 the son of a poor farmer, he entered college late, graduating with honors from Dartmouth in the class of 1791. From 1792 until 1809, he taught at various academies until he returned to Dartmouth as professor of Latin and Greek in 1809. The following year, he was made professor of mathematics and natural philosophy, holding that position until 1833, when he became emeritus.

Although Adams was an active figure aiding Dartmouth through difficult days, especially during the famous Supreme Court Case of 1819, he has not left an impressive record of scientific achievement. Evans's *American Bibliography* (**II,** #32692) indicates that while he was a preceptor at Leicester Academy in Massachusetts he revised and edited an enlarged edition of Nicholas Pike's *System of Arithmetic* (Worcester, Mass., 1797). Outside Dartmouth, Adams's chief interests were in colonization, temperance, foreign missions, and Bible societies.

Chapman, George T. 1867. *Sketches of the Alumni of Dartmouth College* (Cambridge, Mass.), p.57.
NCAB **9**: p.91.

Adrain, Robert (1775–1843). University of Pennsylvania, 1828–1834.

Robert Adrain, born in Ireland, in 1775, escaped to the United States with his family after participating in the Irish Rebellion of 1798. For a decade, he taught mathematics in academies at Princeton, York, and Reading until called to the professorship of mathematics and natural philosophy at Queen's College (Rutgers). By this time, he had already established a reputation, having submitted medal-winning solutions to problems in the New York *Mathematical Correspondent,* founded in the *Analyst* his own short-lived mathematical journal, and introduced Diophantine Analysis to the United States. In 1812 he was elected to the APS.

Remaining three years at Rutgers, Adrain then moved to Columbia College, where he taught for thirteen years. Although he never convinced the Columbia administration to present mathematics at as high a level as that he had reached himself, he had a profound effect upon the College through his formation of a mathematical club. Ill health caused him to retire to his farm near Rutgers in 1827, where he remained until he accepted, a year later, the professorship of mathematics at the University of Pennsylvania. Adrain, serving also as vice-provost, was an extremely unsuccessful administrator; and, too demanding and unpopular with his students, he was embarrassed into resigning in 1834. After this experience, Adrain never returned to college but spent the remaining years of his life teaching grammar school and resting on his farm.

Robert Adrain and Nathaniel Bowditch were the leading mathematicians in America before Benjamin Pierce. Among Adrain's original contributions were a "rule" for correcting dead-reckoning at sea, adopted by Bowditch in the last edition of his *Practical Navigator;* a "rule" for correcting surveying measurements; and a catenary curve, found to have importance a half century later. His greatest achievement—one which would have venerated his name in the history of science had he lived in Europe—was the first proof for what is now known as Gauss's exponential law of error. Adrain also used the law to correct the mean value of the earth's ellipticity as calculated by LaPlace in his celebrated *La Mécanique Céleste* (3). The author of numerous articles, he left even more in manuscript form; and references and locations for these can be found in Cajori (1).

Cajori, Florian. 1890. *The Teaching and History of Mathematics in the United States* (Washington), pp. 66-71.
DAB.

Alexander, Stephen (1806-1883). Princeton University, 1840-1877.

Graduated with honors from Union College, Stephen Alexander taught at Chittenango Academy in New York for six years. In 1830 his sister married their cousin Joseph Henry (*q.v.*), who was teaching at the Albany Academy and was to prove instrumental in forming Alexander's career. While Henry, for instance, was working on electrical problems at Albany, Alexander joined the Academy in some unknown capacity to work in connection with astronomical problems; and his first observations upon star occultations and solar eclipses were transmitted to the Academy in 1831. The following year, when Henry became professor of natural philosophy at Princeton, Alexander traveled there to study in the Theological Seminary. By 1833, he was made mathematics tutor in Princeton College, becoming adjunct professor in 1834 and, in 1840, professor of astronomy, a position he held with some title changes until he retired in 1877.

Although hampered by inadequate telescopes until near his death, Alexander became an expert on star clusters. He traveled to Maryland in 1831 and to Georgia in 1834 to observe eclipses; and, under the auspices of the United States Coast and Geodetic Survey, he headed an expedition to Labrador in 1860 to observe the solar eclipse. In conjunction with Bache (*q.v.*), Espy, and Henry, he determined the difference in longitude between Princeton and Philadelphia by observation of the fall of meteors—the first successful application of Halley's method.

Alexander wrote only one short book *On the Fundamental Principles of Mathematics* (New Haven, 1849), his small output being partially due to the fact that he was a brilliant speaker who only required sketchy notes for his lectures and disdained making them available for the press. His lectures were well received not only by Princeton students, but even by the ladies of the town, who crowded the lecture hall to hear him speak on the nebular hypothesis. For the use of Princeton students he had syllabi printed of lectures on algebra, astronomy, and differential calculus; and he also published a number of articles in the *AJS*, the *APS, Proc.*, and the *Astronomical Journal.* One rather curious essay of his on the "Harmonies of the Solar System" (*Smithsonian Contributions to Knowledge* **21**, no. 2, 1876) is an unsuccessful Keplerian attempt to devise simple ratios for the distances between planets. An original member of the AAAS (1848), Alexander served as president of the Association in 1859. He was elected to the APS in 1839 and was one of the fifty original members of the NAS (1863).

DAB.
Young, C. A. 1886. *NAS Biog. Mem.* 2: pp.249-259.

Allen, William Henry (1808-1882). Dickinson College, 1836-1850.

William Henry Allen attended the Maine Conference Seminary and graduated from Bowdoin College in 1833. For three years he taught classical languages at Methodist Cazenovia Seminary, being appointed to the chair of natural philosophy and chemistry at Dickinson College in 1836. He held this position for ten years and added English literature to his duties in 1846, also serving as acting president of the College from 1847 to 1848. In 1850 he left Dickinson to become president of Girard College in Philadelphia, a position he held until his death in 1882 except for a five-year period from 1862 to 1867 when he served as president of the Pennsylvania Agricultural College.

Active in religious affairs, Allen was president of the American Bible Society for the year of 1872 and authored sermons, school prayer books, and miscellaneous addresses. There exists no further record of his connection with science outside of his tenure at Dickinson.

BDA.
Cleaveland, Nehemiah. 1882. *History of Bowdoin College* . . edited by Alpheus S. Packard (Boston).
1883. *In Memoriam . . . of William H. Allen* (Philadelphia).

Anderson, Henry James (1799–1875). Columbia College, 1825–1843.

A native New Yorker, Henry J. Anderson graduated with honors from Columbia College in 1818 and received a medical degree from the College of Physicians and Surgeons. Spending his spare time studying mathematics, Anderson was appointed professor of mathematics and astronomy at Columbia College in 1825. After he resigned in 1843, he spent the remainder of his life traveling and promoting Catholicism, to which he had converted during a visit to the French astronomer Arago. He also served as geologist on Lynch's expedition to the Dead Sea (1848?) and paid from his own finances to join the American expedition to Australia to observe the transit of Venus in 1869.

Besides the results of his geological studies in the Holy Land, Anderson published one long article, "On the motion of Solids on Surfaces . . . with a Particular Examination of their small oscillatory motions" (APS, Trans., n.s., **3,** 2 (1830): pp.315-382). This essay demonstrates that Anderson was one of the most competent mathematicians of his day, able to solve partial differential equations and to use the results of the French mechanics.

ACAB 1: pp. 68-69.

Bache, Alexander Dallas (1806–1867). University of Pennsylvania, 1828–1836 & 1842–1843.

Alexander Dallas Bache, great-grandson of Benjamin Franklin, was one of the leaders of American science in his day. Bache graduated from the West Point Military Academy in 1825, the youngest and ablest in his class. During three years of military service, he taught at West Point, built fortifications, studied chemistry and physics, and even collected shells.

In 1828 Bache resigned from military service and accepted the position of professor of natural philosophy and chemistry at the then failing University of Pennsylvania. Bache's power of administration quickly emerged as he persuaded the trustees to give proper support to science in the college apart from the Medical School. Leaving the University in 1836, Bache reaccepted appointment in 1842–1843, and, although he did not teach during this second stay, he remained, nevertheless, a long-time friend of both the University and his successor in the chair of natural philosophy.

Bache's scientific ability was quickly realized in Philadelphia by both the APS, which elected him a member in 1829, and the new Franklin Institute, whose research he directed from 1828 to 1836. His major effort at the Franklin Institute was the investigation of steam-boiler explosions, a very important problem to the new industrial economy; and his early contributions to theoretical science included studies of terrestrial magnetism, on which he published significant results even before Gauss made the subject popular in 1837.

Named president of Philadelphia's Girard College in 1836, Bache spent the two years from 1836 to 1838 in Europe, studying school systems there and returning to write his *Report on Education in Europe* (Philadelphia, 1839), which had a profound effect upon the American college. The magnetic observatory he later established at Girard College ran from 1840–1845 and was the first of its kind in the United States.

The last twenty-four years of Bache's life were spent in large-scale government science. In 1843 he left the University of Pennsylvania for the second time to succeed Hassler as superintendent of the United States Coast Survey, helped with the foundation of the Smithsonian Institution, and served as first president of the National Academy of Sciences, which he had helped to found through his leadership of the scientific "Lazaroni." The list of his accomplishments at the Survey, the Institution, and the Academy is considerable. Nathan Reingold of the Smithsonian Institution, who is currently working on a biography of Bache, has called him "the great scientific promoter, organizer, and administrator of his generation—the great tycoon of American science." His scientific publications number over two hundred; and his American edition of Sir David Brewster's *A Treatise on Optics* (Philadelphia, 1838) with an appendix on the analysis of reflection and refraction was used in almost every college for decades.

Bache's illustrious career in research, teaching, and administration cannot be adequately condensed in a few paragraphs. I have utilized in this brief summary my perspective from Philadelphia area manuscript material and especially the biographical essay by Joseph Henry, *Eulogy on Prof. Alexander Dallas Bache* from report of Smithsonian Institution for 1870 (Washington Government Printing Office, 1872). The bibliography which can be found in the eulogy is copied from Benj. A. Gould's address to the AAAS, August 6, 1868 and printed in its *Proceedings* (1869), pp. 1-47.

Daniels, George H. 1968. *American Science in the Age of Jackson* (New York), p. 203, for twentieth-century Bache scholarship.

Baird, Spencer Fullerton (1823-1888) Dickinson College, 1845–1850.

Spencer Fullerton Baird had a great love of nature even as a teenager. While a fifteen-year-old inquisitive student at Dickinson College, he made the acquaintance of the ornithologist Audubon, who was to provide him with numerous species of North American birds after they were sketched. Baird's association with Audubon was only one of the many friendships with both American and European naturalists that enabled him to be so effective an administrator.

After graduating from Dickinson in 1840, Baird went to New York to study natural history on his own and to attend lectures at the College of Physicians. In 1845 he was recalled to his alma mater to become professor of natural history, a title which was changed in 1846 to professor of natural history and chemistry. In his brief five-year stay at Dickinson, Baird assembled a notable natural history collection for the college and succeeded in stimulating a number of future naturalists. In 1848 he received from the newly formed Smithsonian Institution a modest grant—its first grant for original research.

By 1850, he had come to the attention of Joseph Henry, who personally invited him to become assistant secretary of the Smithsonian Institution, where he remained for the duration of his life, becoming secretary after Henry's death in 1878. The collections which he assembled still form the nucleus of the Institution's Natural History Museum.

Almost all of Baird's prodigious labor at collection and classification was financed by the government. His first important work, which became known as *Birds of North America* (1858), was compiled in conjunction with a survey of the western railroad routes. By the time this pioneer taxidermist died, he had authored 79 articles on ornithology and over 125 articles on other topics of natural history, in addition to having had some connection with nearly a thousand publications principally written by others. Reflecting his interest in the life cycles and preservation of fish, Baird agreed to head the United States Commission of Fish and Fisheries in 1871. His long, varied, and continuously active career makes Baird one of the most influential of nineteenth-century American naturalists.

DAB.
Goode, George Brown. 1883. *The Published Writings of Spencer Fullerton Baird, 1843-1842, Bulletin of the United States National Museum* (Washington) **20:** 377 pp. (The book contains a biography and extensive bibliography).
1890. *Smithsonian Institution Annual Report,* July 1888. pt. 1 (Washington), pp.703-744 (various biographical memoirs).

Benedict, Farrand Northrup (1803-1880). University of Vermont, 1833-1854.

A native of Parsippany, New Jersey, Farrand N. Benedict graduated with highest honors from Hamilton College in 1823. For the next decade, he studied law, taught (presumably in academies), and was a practicing civil engineer. In 1833 Benedict became professor of mathematics and civil engineering in the University of Vermont, a position he held until he resigned to care for his ailing wife in 1854.

Benedict, a frequent contributor to current mathematical publications, had a keen and original mathematical mind. In fact, although students liked him, they often had difficulty following his long excursions from the prescribed textbooks.

Outside of college, Benedict made a number of explorations and surveys of the Adirondack Mountains. From 1837 to 1838 he was assistant to Ebenezer Emmons (*q.v.*) on the geological survey of New York; and the Royal Society has meteorological observations made at Burlington by Benedict and his faculty colleague and cousin, G. W. Benedict (*q.v.*). After his retirement, he made a hydrographic survey of the state of New York, a report of which was published in 1875; and he was consulted on the construction of reservoirs. Benedict joined the AAAS at its inception in 1848.

Thomas, Evan. 1926. " 'Little Ben' Mathematical Genius." *Vermont Alumni Weekly* 5: pp. 391-392. (Copy of this article and additional information supplied by present Vermont archivist, T. D. Seymour Bassett.)

Benedict, George Wyllys (1796-1871). University of Vermont, 1825-1847.

A graduate of Williams College, class of 1818, Benedict taught in secondary academies for three years after graduation and returned to Williams as instructor from 1821 to 1825. Upon the resignation of James Dean, the University of

Vermont's first professor of mathematics and natural philosophy, Benedict was chosen to fill the position in 1825. As the University grew in size and number of professorships, Benedict's title was changed to professor of natural philosophy and chemistry (1829–1839) and to professor of natural history and chemistry (1839–1847).

Always active in university affairs, Benedict sponsored the "College of Natural History" at the University in 1826 and was the principal architect of the faculty report of 1827, *Exposition of the System of Instruction and Discipline pursued in the University of Vermont* (Burlington, 1829), which President Wayland of Brown called "by far the best thing on college education that I had seen." One of the immediate results of this report was Vermont President James Marsh's new plan of 1829, by which a separate professor of mathematics and civil engineering was elected, later becoming the first to offer the calculus there.

Although Benedict loved science and often lectured to the citizens of Burlington on popular scientific subjects, particularly electricity, he had no scientific publications. The Royal Society of London, however, possesses some straightforward meteorological data which Benedict and his colleague Farrand N. Benedict (*q.v.*) forwarded to Sir J. F. W. Herschel in 1839 (MSS, APS, film). These observations for Burlington on the day of the vernal equinox reflect the period's widespread interest in the determination of weather patterns.

In the 1840's, Benedict became interested in the electric telegraph and the United States Coast Survey's novel applications of this device in the determination of exact longitudes. He reported to the Vermont trustees in 1847 that they had failed to honor the other colleges' commitments to science and resigned to enter the more lucrative telegraph industry. He first worked for Ezra Cornell, the founder of Cornell University, in the construction of a telegraph line from Troy to Canada junction. Benedict then joined the Vermont and Boston telegraph company, which constructed between White River Junction and St. Johnsbury a line that later joined Burlington to Boston.

BDA.
Lindsay, Julian Ira. 1954. *Tradition Looks Forward* (Burlington, Vt.).
MSS, University of Vermont and APS.

Caswell, Alexis (1799–1877) Brown University, 1828–1863.

Educated in his early years at Taunton Academy, Alexis Caswell graduated at the head of his class at Brown in 1822. For five years he taught at the Columbian College in Washington while studying theology and, in 1827, became a Baptist minister in Nova Scotia. A year later, he left to become professor of mathematics and natural philosophy at his alma mater. In 1851 he was made professor of mathematics and physical astronomy, although astronomy was always his special love. Caswell was a thorough and competent teacher and an active correspondent with other scientists. A member of numerous societies, he served as vice-president of the AAAS in 1850 and was selected by Congress as one of the original fifty members of the NAS in 1863.

After his retirement from teaching in 1863 because of poor health, Caswell entered finance and industry for five years until Brown begged him to return as president, a position he filled admirably for the four years from 1868 to 1872. The remainder of his life he spent working in civic concerns and revising his meteorological observations. Thoroughly conversant with the astronomy of his day, Caswell was often called upon to review books. His own writings include these reviews, a few memorial addresses for other scientists, a survey of American achievements in astronomy (AAAS Presidential Address, Aug., 1859 in *Proc.* **13:** pp.1–26), and his "Meteorological Observations," made at Providence, Rhode Island, 1831–1860, and published as *Smithsonian Contributions to Knowledge* **12** (Washington, Smithsonian Institution, 1860). These tables were later extended, condensed, analyzed, and published as "Results of Meteorological Observations . . ." (**24,** same series, 1885).

In conclusion, it is interesting to note Caswell's stance in the evolutionary controversy as revealed in the first part of his presidential address to the AAAS (*Proc.,* **13** (1859): p.9): "I shall not hesitate, in this presence, to give utterance to my own profound conviction that all true science is in harmony with the Bible, rightly interpreted."

Gammell, William. 1877. *Memoir of Alexis Caswell . . .* (Boston). (Bibliography on p.9.)
DAB.

Chace, George Ide (1808–1885). Brown University, 1834–1872.

Son of a Baptist farmer in Lancaster, Massachusetts, George I. Chace attended the Lancaster Academy, entered the sophomore class of Brown University in 1827, and graduated with first honors in 1830. He then taught in the Waterville, Maine, Academy for a year until he was appointed mathematics tutor at Brown in 1831. At this time he traveled to the University of Pennsylvania to attend the celebrated Robert Hare's lectures on chemistry, anatomy, and physiology; by 1833, he had advanced at Brown to the position of adjunct professor of mathematics and natural philosophy. He became, in 1834, professor of chemistry at Brown; and, in 1836, his chair was expanded to include geology and physiology, subjects he taught until 1867. Chace gained experience in geology in 1836 during a trip to Kentucky and Virginia when he began his mineral collection, which became the Brown geological collection, supervised for many years by Chace and augmented by exchange with other scientists.

Chace's lectures at Brown were described by James Angell, later president of the University of Michigan, as "in themselves almost demonstrations" of the sciences. He also lectured to the citizens of Providence at the high school and was called upon by local manufacturers for advice on chemical processes, especially by the industrialists of the gold mine fever days of 1863–1866. Among his close personal friends he numbered R. Hare, A. Guyot, J. D. Dana, and J. Henry, who often urged him to publish some of his scientific results—a suggestion which, in his extreme modesty, Chace always refused to adopt. His publications, therefore, include only one pamphlet and about a dozen articles on metaphysics and natural theology published in *Bibliotheca Sacra* between 1848 and 1850. Though a deeply religious man who sincerely believed that God's moral purposes could be seen in the design of the physical world, Chace flatly rejected the view that God, the "first cause," superintends every "second cause," or process of the purely physical world.

BDA.

Chace, George I. 1854. *The Relation of Divine Providence to Physical Laws* (Boston).

Murray, James O., editor. 1886. *George Ide Chace, L.L.D., A Memorial* (Cambridge, Mass.). (p. 70 contains Chace's bibliography).

Chase, Stephen (1813-1851). Dartmouth College, 1838-1851.

After graduating from Dartmouth College in 1832, Stephen Chase studied theology at Andover Seminary for two years. From 1834 to 1838, he taught in various academies as far away as Maryland and Virginia, returning to Dartmouth in 1838 to become professor of mathematics until his death in 1851. In 1849 he published *A Treatise on Algebra* (New York, 1849), which was used at Dartmouth for decades as well as at a few other colleges. Apparently a brilliant thinker and teacher, Chace, had he lived past the age of thirty-seven, might have left a greater legacy than one book.

Chapman, George T. 1867. *Sketches of the Alumni of Dartmouth College* (Cambridge, Mass.), p.261.

Smith, Baxter Perry. 1878. *History of Dartmouth College* (Boston), pp. 298-303.

Cleaveland, Parker (1780-1858). Bowdoin College, 1805-1858.

Son of a Massachusetts physician, Parker Cleaveland attended Dummer Academy and graduated from Harvard College in 1799. While holding teaching jobs in Massachusetts and later in Maine, Cleaveland spent three years reading law, but in 1803 decided to study theology instead. He quickly deserted his few months of ministerial preparation to become tutor of mathematics and natural philosophy at Harvard; in 1805, he accepted the position of professor of mathematics and natural philosophy at newly formed Bowdoin College, where he remained until his death in 1858.

After two years at Bowdoin, Cleaveland was contacted by a lumbering company in Maine which had just excavated for a sluice route and uncovered some rocks that looked as if they might be valuable. At that time, the science of mineralogy had not yet reached America, and the professor had some difficulty in making the identifications. The problem, however, set him about reading all available works of the European mineralogists and beginning his own collection of local minerals, which by acquisition and exchange was to become the college's "mineralogical cabinet." In the course of building up his collection, Cleaveland published a few articles in the *Memoirs* of the APS; but it was not until 1816 that he published his *Elementary Treatise on Mineralogy and Geology* (Boston, 1816), the first American text on mineralogy.

Cleaveland's *Treatise* immediately drew world-wide attention, not only because it was the first systematic arrangement of a previously unexplored branch of American science, but also because it so masterfully applied the combined French and German classification system first suggested by Brongniart. He received letters of praise from Brewster, Davy, Berzelius, Cuvier, Häuy, Goethe, and Brongniart, with most of whom he conducted long years of correspondence. Elected to the APS in 1818, Cleaveland also secured membership in a score of European societies. The Geological Society of London arranged its own collections according to Cleaveland's format; and seven American colleges tried unsuccessfully to wrest him from Bowdoin.

For such a popular work, the *Treatise* surprisingly went through only a second edition (1822) and was then quickly replaced in the American college by newer books. Cleaveland was far more interested in teaching than in publishing; and the material he collected for a third edition of his book still remains in manuscript. Yet as a teacher at Bowdoin he shall ever be known. Living to the age of seventy-eight, he was one of the last men able to instruct college classes in almost all of their sciences. But although he was originally appointed in 1805 as professor of mathematics and natural philosophy, he dropped mathematics entirely in 1828 to become professor of chemistry, mineralogy, and natural philosophy. Over the years, some of the natural philosophy was assumed by other professors, and he was left to teach only natural history and chemistry.

Beyond his teaching duties, Cleaveland did a limited amount of popular lecturing in cities of Maine and New Hampshire, transporting his chemical equipment by oxcart. In addition, President Van Buren appointed him commissioner for the survey of the North Eastern Boundary; and President Pierce, one of his former students, appointed him a regent of the Smithsonian Institution.

DAB.
Woods, Leonard. 1860. *Address on the Life and Character of Parker Cleaveland* (2nd ed., Brunswick, Maine).
Extensive manuscript collection at Bowdoin College.

Cooke, Josiah Parsons (1827–1894). Harvard College, 1850–1894.

Son of a wealthy Boston lawyer, Josiah P. Cooke attended the Boston Latin School and graduated from Harvard College in 1848. While still a boy, he had been stimulated by Benjamin Silliman's (*q.v.*) lectures in Boston; and he now traveled to Europe to further his education at the chemical laboratories of Regnault and Dumas. On his return in 1849, he was made tutor at Harvard and the following year was named Erving Professor of Chemistry and Mineralogy, holding that position until his death in 1894.

At Harvard, Cooke campaigned for more science in the undergraduate program. Within a decade, he had established general chemistry as a requirement for all Harvard students and analytical chemistry as an additional elective. In addition, he was a pioneer in the use of the laboratory as a means not only of demonstration but of actual student participation.

Cooke was an eminent chemist who participated in many scientific activities, including popular lecturing in many cities from Boston to Washington. A specialist in problems of the periodical table, he conducted his most acclaimed research on the determination of the atomic weight of antimony. His first full-length textbook, *Elements of Chemical Physics* (Boston, 1860; 2nd ed., 1873), was the first in this country to employ a molecular conceptual basis of chemical science and was even translated into most of the languages of Europe. Cooke performed many difficult mineral analyses, publishing the results in the *AJS* and the *AAAS, Proc.*, both of which journals he served as editor. Many of his papers are collected in *Contributions to Chemistry and Mineralogy from the Laboratory of Harvard College* (New Haven, 1867) and *Contributions from the Chemical Laboratory of Harvard College* (Cambridge, 1877). In all, Cooke published 41 scientific papers, 32 papers on other subjects, and 7 books on chemistry proper, in addition to a curious volume entitled *Religion and Chemistry* (New York, 1864) in which he extends the doctrine of "design" into the constitution of the atmosphere and inorganic matter.

Elected to the NAS in 1872, Cooke was a member of the Royal Institution and became one of the few nineteenth-century Americans to be elected to the Chemical Society of London.

ACAB 1: p. 721.

Jackson, Charles Loring. 1902. *NAS Biog. Mem.* 4: pp. 175–183. (contains bibliography).

Dana, James Dwight (1813-1895). Yale College, 1850–1890.

James Dwight Dana was born and raised in Utica, where the high school was operated by Union College graduates who were careful to include scientific instruction in the curriculum. About 1829–1830, in fact, Dana may well have been a student of Asa Gray (*q.v.*), who was then teaching at the high school. Attracted, thereafter, to Yale College because of the fame of the elder Silliman (*q.v.*), Dana graduated from that school in 1833, taking his first job as schoolteacher on board a naval vessel. He utilized this opportunity of free world travel to make mineralogical and geological observations on the places he visited; and his first paper, "On the Condition of Vesuvius in July, in 1834" (*AJS,* **27** [o.s.], 1835, pp.281-288), was sent from Smyrna. His second paper, "A New Mineralogical Nomenclature" (*ibid.* **28** (1835): pp.250–262), was a remarkable work for a young man, since he prepared in it a mathematical representation of mineral classes which superceded the older descriptive nomenclature. This article, later incorporated into his *System of Mineralogy* [and crystallography] (New Haven, 1837), was acclaimed throughout the world; and Dana's *Manual of Mineralogy* as revised by Cornelius S. Hurlbut is now, in its seventeenth edition (New York, Wiley, 1959), a permanent memorial to his work.

In 1836, Dana succeeded C. U. Shepard (*q.v.*), Oliver P. Hubbard (*q.v.*), and B. Silliman, Jr. (*q.v.*) as laboratory assistant to Silliman, Sr.; in 1838, commissioned as mineralogist for the voyage, he set sail with the Navy's famous Wilkes Expedition. Much of the work he accomplished on this four-year voyage, however, was in geology and zoology, owing to the retirement of the naturalist. On his return in 1842, Dana spent thirteen years writing the results of his researches, the three reports filling 3,117 pages. By 1854, he had completed revisions of his two books totaling 574 pages, and 64 articles totaling 832 pages. Married to one of Benjamin Silliman's daughters, Dana became an editor of *Silliman's Journal* in 1846.

Yale, in 1850, appointed Dana professor of natural history, but he did not regularly discharge his duties until 1855. Later (1864), his title was changed to professor of geology and mineralogy, a position which he held until he stopped teaching in 1890. Dana's fantastic work schedule resulted in a nervous breakdown in 1859, from which he never totally recovered. Nevertheless, in the last thirty-five years of his life, he still managed to contribute dozens of papers on geology, mineralogy, and zoology.

Dana was not a brilliant teacher; and his efforts at Yale were not fully appreciated until he taught specialized advanced courses in the latter part of the century. A member of all major scientific societies, president of the AAAS in 1855, and one of the fifty founding memers of the NAS (1863), Dana received honors from numerous foreign societies, including the Geological Society of London and the Royal Society.

Persson, Louis, 1919. *NAS Biog. Mem.* **9**: pp.40–42 (contains a bibliography of Dana's works and of biographical memoirs).

Dod, Albert B. (1805–1845). Princeton University, 1830–1845.

Albert Dod graduated from Princeton in 1822 and worked as a private tutor in Virginia for three years. He then returned to Princeton Theological Seminary and was ordained in 1828, spending the remainder of his life in teaching and preaching. In 1830 Dod was made professor of mathematics at Princeton, his primary association until his early death in 1845. He was, however, a frequent lecturer and contributor to journals on such topics as capital punishment, phrenology, transcendentalism, theology, and architecture—a subject which Joseph Henry (*q.v.*) persuaded him to offer at Princeton. One of Dod's most interesting reviews was of Robert Chamber's controversial *Vestiges of the Natural History of Creation* (1844). Here, after a careful nonpolemical destruction of the *Vestiges* on scientific grounds alone, he also voiced his objection "against [the] derivation of man from the brutes."

Dod was a brilliant but not original mathematician who gained a reputation at Princeton for overemphasizing mathematics. The best evidence that he knew his subject can be found in his own critique of analytical geometry texts (essay, 1841) and in Cajori's statement that Dod helped prepare the American edition of Young's *Elements of the Differential Calculus* (1833), one of the earliest calculus texts to appear in this country.

DAB.

Cajori, Florian. 1890. *Teaching and History of Mathematics in the United States* (Washington), p. 395.

Dod, Albert B. 1847. *Essays, Theological and Miscellaneous . . . Including Contributions of the late Rev. Albert B. Dodd (sic)* (New York). (contains seven of Dod's essays, including those on analytical geometry and the *Vestiges*).

Emmons, Ebenezer (1799–1863). Williams College, 1833–1854.

While still a student at Williams College from 1816 to 1818, Ebenezer Emmons increased his love of nature through contact with Amos Eaton. He also received a degree from Eaton's Rensselaer Institute in 1826; and to complete his formal education, he earned a medical degree in the Berkshire Medical School. In 1828 he gave lectures in chemistry at Williams while maintaining a private practice; by 1833, he had been appointed professor of natural history at Williams. This position he held until 1854, although his tenure was more than a few times interrupted by his wide range of outside activities. After 1838, he commuted to Williamstown from Albany for a few months a year to give his lectures, then principally in mineralogy and geology.

Emmons joined the new State Geological Survey of New York in 1836, moved to Albany in 1838, and held the position of professor of obstetrics at Albany College until 1852. His first major work, the *Geology of New York. Part II. Comprising the Survey of the Second Geological District* (Albany, 1842), was published in 1842. Prior to this, he had published a few articles in medical journals and the section on quadrupeds in *Reports on the Herbaceous Plants and the Quadrupeds of Massachusetts* (Cambridge, 1840). From 1842 until 1854, he was involved in the enormous task of producing an agricultural survey of New York State, the results of which were published in six (actually five, with volume 3 in two parts) huge and beautifully illustrated quarto volumes under the title of *Agriculture of New York* (Albany, C. Van Benthuysen, 1846–1854). The last volume of this series, on entomology, was of obvious aid to the practicing farmer.

After his New York studies, Emmons moved south in 1854 to become state geologist of North Carolina, where he died in 1863. His last publications were *American Geology* (Albany, 1856) and a *Manual of Geology* (Phila., 1860). Although a prodigious worker and writer, Emmons was disliked by students for his fuzzy lectures and by fellow scientists for the unethical manner in which he failed to give proper credit to others for their work. Nevertheless, even if guilty of plagiarism, he contributed much that was, in fact, original. His discovery of a geological layer that he termed the "Taconic" during the New York survey caused a debate which found no resolution in the geological world until the 1880's; and it remains to his credit that Emmons was one of the founders of the 1840 Association of American Geologists, which became, in 1848, the AAAS.

DAB.

Perry, Arthur Latham. 1899. *Williamstown and Williams College* (Williamstown, Mass.), pp. 517-523.

Farrar, John (1779-1853). Harvard College, 1807-1836.

John Farrar, son of a Massachusetts farmer, attended Philips Academy and graduated from Harvard College at the age of twenty-four in 1803. Farrar studied theology at Andover Seminary until he returned to Harvard as Greek tutor in 1805. When the Harvard Corporation could not find an established scientist to assume the Hollis Professorship of Mathematics and Natural Philosophy in 1807, it chose Farrar to fill the position on the basis of his interest in the subjects. For three decades, this brilliant lecturer and energetic writer served Harvard; and, although he was only a mediocre scientist, he left his mark upon the college course.

Farrar's astronomical and meteorological observations appeared in the *North American Review* and the *Transactions* of the American Academy of Arts and Sciences, where he served as officer from 1811-1831. The author of a dozen books, Farrar introduced French mathematics into the American college. Beginning with his translation of Lacroix's *Arithmetic* in 1818, he prepared English translations of the algebra, geometry, trigonometry, calculus, optics, mechanics, and astronomy of such recent French scientists as Biot, Arago, and Legendre. His own *First Principles of Differential and Integral Calculus* (1824) was the first American textbook to employ the notation of Leibnitz.

With the exception of a translation of Arago's *Tract on Comets* (1832) and revisions of earlier works, all of Farrar's writings were completed in the decade from 1818 to 1828. Known together as the "Cambridge Course of Mathematics and Mechanics," Farrar's works were adopted by most colleges. Although soon replaced by more sophisticated works, the "Cambridge Course" was the first curriculum which provided American college students with a sense of contemporary science.

Poor health caused Farrar to resign in 1836; although he lived until 1853, his last few decades, spent in much suffering yielded no scientific activity.

DAB. (Contains partial bibliography of books.)

Cajori, Florian. 1890. *Teaching and History of Mathematics in the United States* (Washington), pp. 129–133.

Palfrey, John Gorham. 1853. *Notice of Professor Farrar* (Boston). (Contains partial bibliography of articles, p. 10.)

Foster, John (1811–1897). Union College, 1836–1897. (?).

Graduating from Union College in 1835, John Foster spent his long life teaching either mathematics and natural philosophy or natural philosophy alone at his alma mater. A member of the AAAS from its beginning in 1848, Foster wrote an extremely carefully arranged volume in *An Elementary Lecture on Electricity, Magnetism, Galvanism, Electro-Magnetism and Acoustics* (Schenectady, 1877), which, in 1884 was displayed at the International Electrical Exhibition.

Union College archivist Mrs. Frances Miller informs me that there is no record that Foster had any associations outside of Union College. In his *Electricity* (p.84), there is a suggestion that Foster also wrote a textbook on dynamics, which I have been unable to find.

Fowler, William Chauncey (1793–1881). Middlebury College, 1828–1838.

A graduate of Yale College, class of 1816, William C. Fowler studied at the Yale Divinity School while acting as rector in New Haven's Hopkins Grammar School. He served as a tutor at Yale from 1819 to 1823; pastor of a Congregational church, 1825–1827; professor of chemistry and natural history at Middlebury College, 1828–1838; and professor of rhetoric and literature at Amherst College, 1838–1843. The remainder of his long life was spent in politics and in the authoring of numerous historical and literary works, including a university edition of *Webster's Dictionary* (1845). Of these fifty-nine professors, Fowler is one of the handful who seem to have had no aspects to their professional scientific lives other than a few years of college teaching.

BDA.

Frazer, John Fries (1812-1872). University of Pennsylvania, 1844-1872.

John F. Frazer, a Philadelphian, attended public schools in that city and private school in Connecticut before coming to the University of Pennsylvania. Graduating in the class of 1830, he stayed on at the university as Alexander D. Bache's (*q.v.*) laboratory assistant, receiving the A.M. in 1833. For the next few years, he studied both medicine and law, and was admitted to the Bar. During the summer of 1836, he worked as assistant to Henry D. Rogers (*q.v.*) on the first geological survey of Pennsylvania and then, from 1839 to 1844, taught science at the new Central High School in Philadelphia. During this time, Frazer assisted Bache in the first Amerierican measurements of the effect of the aurora borealis upon the earth's average magnetic intensity and the systematization of the secular variations.

Known for his critical evaluation of the work of others rather than for his own contributions, Frazer was, nevertheless, a prominent figure in the life of American science. For many years editor of the *Journal of the Franklin Institute,* he organized many of its important researches on steam-boiler explosions with Bache. Later, when he taught in college, he always paid special attention to steam problems.

When Bache left the University of Pennsylvania for the last time in 1843, he personally nominated Frazer to succeed him; from 1844 until his death in 1872, Frazer occupied the chair of natural philosophy and chemistry. In addition to serving as vice-provost from 1855 to 1868, Frazer was at the center of many forward movements at the University, including the formation of a department of mines, arts, and manufactures (MSS, U. of P., May 20, 1848), the plan for graduate education (1852), and the preparation for the new scientific laboratory (1872).

In addition to science, Frazer was interested in the history of science and literature. He was elected in 1842 to the APS, where he acted on many committees to review works for publication, besides serving as secretary of the society in 1845 and vice-president during the years 1855-1858 and 1867. Frazer was also a member of the AAAS and one of the fifty original members of the NAS (1863).

DAB.
APS Archives.
Cheyney, Edward Potts. 1940. *History of the University of Pennsylvania* (Philadelphia), pp. 238-240.

Gillespie, William Mitchell (1816–1868). Union College, 1845–1868.

After graduating in 1834 from Columbia College where he had studied under James Renwick (*q.v.*), Gillespie spent a decade as traveler, student, and engineer, furthering his engineering education at the Ecole des Ponts et Chaussées. Upon his return to the States, Gillespie was appointed professor of civil engineering in Union College, a title later changed to professor of civil engineering and adjunct professor of mathematics "to be serviceable in the title pages of the scientific books which I have in progress." According to one student, Gillespie "kept fully abreast with all the latest developments in engineering science. He was familiar with everything published in English or French, and would bring into the classroom his entire fund of knowledge." His personal library contained almost 800 volumes of mathematics, science, and engineering texts without any works on natural history.

Gillespie helped shape the American engineering curriculum at both Union and other colleges. In addition to working elective and required courses in engineering into the college program, Gillespie soon established a full four-year engineering option. Among his own works were *The Philosophy of Mathematics* (1851)—a translation of part of August Comte's *Cours de Philosophie Positiv, A Manual of the Principles and Practice of Roadmaking* (1847), *Treatise on Land-Surveying* (1855; privately printed, 1851), and *Treatise on Leveling, Topography, and Higher Surveying* (posthumous, 1870). His first treatise on surveying had gone through ten editions by 1873; and the treatise on higher surveying was revised as late as 1906 and still used as a textbook in the twentieth century.

DAB.
1906. *Union College, Class of 1856 Book* (Schenectady), p. 24. (Copy in Union College Archives.)

Gray, Asa (1810–1888). Harvard College, 1842–1873.

Asa Gray, son of a New York tanner turned farmer, attended Clinton Grammar School and Fairfield Academy, studied with Albany area doctors and naturalists, and received a medical degree from the College of Physicians and Surgeons of the Western District of the State of New York (Fairfield) in 1831.

Already interested in plants for years, Gray was corresponding with Lewis Beck and John Torrey (*q.v.*) by 1830. For a decade he sought means of support while pursuing his passionate interest in plant identification. The doctor, who rarely practiced, did, however, teach at Utica Gymnasium (1832), Hamilton College (summer, 1836), and even at Sunday school (1835). In 1835 Gray moved to New York to be near Torrey, with whom he collaborated on the monumental but unfinished *Flora of North America.* Soon his *Elements of Botany* (New York, 1836) was published—the first full-scale attempt to use the newer natural system of botanical classification in America. In addition, Gray became librarian of the New York Lyceum of Natural History, one of the few paying jobs available with the scientific societies. In 1838 Gray was appointed to the first exclusively botanical chair in America at the University of Michigan; and, although he resigned his appointment to the Wilkes Expedition because of the many delays, the Michigan appointment gave him an opportunity to travel to Europe to buy books, visit naturalists, and inspect American specimens which had been collected in past centuries and deposited in European botanic gardens. While in Europe, Gray also met the young Charles Darwin.

Without ever having been to Boston, Gray was chosen Fisher Professor of Natural History at Harvard, entering upon his duties in 1842. At Harvard, he taught only in the College but had a primary responsibility for the care of the neglected botanic garden there, being by now the premier botanist in America. Gray soon provided evidence on plant geographical distribution that was to prove so indispensable to Darwin that Gray would be honored with receipt of one of the three advance copies of the *Origin of the Species* in 1859.

A long time editor of the *AJS,* Gray was in a unique position to support Darwin in America through his reviews and notices. Although known quite properly as a staunch supporter of evolution, Gray saw a problem in calling natural selection the cause of variation. Thus, he utilized traditional beliefs in the power of Divine Agency to cause and supply regular direction to evolutionary change.

Going into retirement in 1873, Gray remained active in the scientific world until his death in 1888. A founding member of the AAAS (1848) and the NAS (1863), Gray not only belonged to all important American societies but was awarded honorary degrees by Oxford, Cambridge, and Aberdeen.

DAB.
Farlow, W.G. 1895. *NAS Biog. Mem.* **3:** pp. 161–175.
1886. *AJS* **36:** 42-page appendix is a bibliography of all of Gray's works, reviews, notices, etc.
Dupree, A. Hunter. 1959. *Asa Gray* (Cambridge, Mass.) is one of the finest scientific biographies and the principal source for this sketch.

Hackley, Charles W. (1809–1861). Columbia College, 1843–1861.

Charles W. Hackley of New York State graduated in 1829 from the United States Military Academy where, as second lieutenant in the artillery, he taught mathematics at times between 1829 and 1833. In that year he resigned and moved to the University of the City of New York, where he taught mathematics until 1838. While in New York, he studied law and theology, being admitted to the orders in the Protestant Episcopal Church in 1835. From 1838 to 1843, Hackley spent a year as president of Jefferson College in Mississippi and served as a church official. Named professor of mathematics and astronomy in Columbia College in 1843, he remained at Columbia until his death in 1861, joining the AAAS in 1851.

In addition to his contributions to journals, Hackley authored five well-written textbooks: *Treatise on Algebra* (New York, 1846); *Elementary Course of Geometry* (New York, 1847); *School Algebra* (New York, 1848); *Elements of Trigonometry* (New York, 1850); and *Treatise on Trigonometry* (New York, 1853). His two treatises on algebra and trigonometry deserve special notice because both were novel attempts to expand greatly the presentations, especially of proofs, for college students. *The Treatise on Trigonometry* was one of the earliest works to derive the formulae, while the *Treatise on Algebra* was the first American book to include Diophantine analysis. *Silliman's Journal* remarked that Hackley "has presented the American public with the most complete treatise on algebra in the English language."

BDA.

Henry, Joseph (1797–1878). Princeton University, 1832–1846.

The exceptional mental abilities of Albany's Joseph Henry were visible at an early age. After graduation from the Albany Academy, Henry became in 1815 a private tutor to the Van Rensselaer family, where in his odd hours he read such volumes as LaGrange's *Mécanique Analytique*. In the early twenties, Henry considered becoming a medical doctor; but, while he was serving as assistant on a surveying party for the Erie Canal in western New York, his interests were redirected toward science. In 1826, Henry became professor of mathematics and natural philosophy at the Albany Academy, beginning in his spare time a series of electrical researches which was to number him among the world's most original scientists. In this short biography, it is impossible to recount the sequences or the disputed priorities attached to Henry's electrical discoveries—the "intensity" electro-magnet so necessary for the development of the telegraph, the induced current, or the oscillatory nature of the electric discharge; but it is enough to note that his genius was eventually recognized on both sides of the ocean.

In addition to his studies on electricity, Henry also conducted research on solar radiation, on the capillarity and cohesion of liquids, and on terrestrial magnetism. As Simon Newcomb remarked near the beginning of his memoir, "Few have any conception of the breadth of the field occupied by Professor Henry's researches, or of the number of scientific enterprises of which he was either the originator or the effective supporter."

Of greatest interest to us is the fact that out of about 150 papers authored by Henry, 46, including the vast majority of his sustained electrical researches, were published between 1832 and 1846 when he was actively engaged as professor of natural philosophy at Princeton. Henry's only book other than compilations of papers, *Syllabus of Lectures on Physics* (Princeton, 1844), was prepared for the students at Princeton College, and, except for the year 1837 when he was on leave to study in Europe, Henry faithfully instructed all of the classes in natural philosophy, providing extra excitement for many students with the little model of the telegraph which he invented.

Henry did not officially sever his connection with Princeton College until 1848, although in 1846 he left to become first secretary and director of the Smithsonian Institution. Here he exerted great influence in determining the Institution's primarily research-supporting orientation. In a few years, Henry no longer had the time to devote to long projects; and his publications reflect much of the government's scientific interests—lighthouses, meteorology, lightning-rods, hygrometry, and even the heating of the Capitol. Naturally, Henry belonged to every well-known society—the APS (1835), the AAAS (founding member, 1848), and the NAS (founding member, 1863).

DAB.
Newcomb, Simon. 1905. *NAS Biog. Mem.* 5: pp. 1-43.
Daniels, George H. 1968. *American Science in the Age of Jackson* (New York), pp. 211-212. (Contains a bibliography of bibliographies and works in progress).
Reingold, Nathan, editor. 1972- . *The Papers of Joseph Henry* (Washington).

Hitchcock, Edward (1793-1864). Amherst College, 1825-1864.

Edward Hitchcock was born into an orthodox Calvinist family and, except for a youthful flirtation with Unitarianism, remained orthodox throughout his life. Primarily a self-taught scientist, he was inspired by the public lectures of Amos Eaton and worked in the laboratory of Benjamin Silliman while a student in Yale's theology department.

After receiving the degree of Doctor of Divinity from Yale, Hitchcock was called in 1821 to the ministry of Conway, Massachusetts, resigning in 1825 because of "ill health" that was more likely hypochondria. He returned to New Haven to attend Silliman's lectures in preparation for a teaching career at the newly established Amherst College, where he remained until his death in 1864, variously holding the title of professor of natural history and chemistry (1825-1845), president (1845-1854), and professor of natural theology and geology (1845-1864).

By 1835, Hitchcock had become a well-known geologist. He won a national reputation for his work as head of the state geological survey of Massachusetts; and by 1840, his analysis of gorge formation, glacial drift, and stratigraphy had secured him international fame. Indeed, on his trip to the United States, Sir Charles Lyell traveled out of his way to visit Amherst in order to see Hitchcock's most famous find, the "fossil bird footprints," or dinosaur tracks. A corresponding member of the Geological Society of London, Hitchcock, at the invitation of Sir David Brewster, addressed the 1850 meeting of the British Association held in Edinburgh on the topic of stratigraphy, and one of his additional purposes in traveling to Europe in 1850 was to secure information on the operation of agricultural schools for the state of Massachusetts. Not only a member of the APS and numerous other societies, Hitchcock also helped to establish the Association of American Geologists in 1840, becoming its first president. When the society expanded to become the AAAS in 1848, Hitchcock became a member of the Standing Committee; and he was elected by Congress to the NAS at its founding in 1863.

Hitchcock published seventy scientific papers and fourteen scientific books, including his *Elementary Geology* (New York, 1840), which went through thirty-one editions between 1840 and its revision in 1860. In addition, his religious writings include thirty-seven essays and five books. He is best remembered in the twentieth century as a geologist of the first rank; yet in mid-nineteenth-century America, he was perhaps most widely known as the reconciler of science and orthodoxy. Despite his world-wide scientific eminence, Hitchcock spent his entire professional life as teacher at Amherst College, which still displays his impressive geological collections of the Connecticut Valley.

DAB.
Lesley, J.P. 1877. *NAS Biog. Mem.* **1:** pp. 113–134.
1895. *American Geologist* **16:** pp. 133–149. (Contains a bibliography of Hitchcock's works).
Extensive manuscript collection, Amherst College Library.

Hopkins, Albert (1807–1872). Williams College, 1829–1872.

Albert Hopkins, brother of William's famous president and educator Mark, graduated from Williams in 1826. The next three years he spent as tutor at Williams while studying agriculture and engineering (probably at R.P.I.). In 1829, he was appointed professor of mathematics and natural philosophy at Williams; but because he was not a particularly gifted mathematician, that part of his duties was dropped in 1838.

Hopkins traveled to Europe in 1834 to procure scientific equipment for the College. On his return, he succeeded in interesting Williams in an astronomical observatory, which was started in 1835, and in a magnetic observatory, begun in 1837. He organized in 1835 the first extended expedition for college students when he took a group of Williams students to Nova Scotia; and later he established a natural history club.

Outside of a few months' association with the New York Survey in 1836, Hopkins seems to have spent his entire adult life at Williams. There is no record of his authorship of scientific publications; and there is evidence that his lecturing, though effective, was too didactic, proving inferior to instruction given elsewhere. Hopkin's publications consist only of an address at the opening of the Williams Observatory (1838), a sermon, and a library catalogue.

BDA.
Perry, Arthur Latham. 1899. *Williamstown and Williams College* (Williamstown, Mass.)

Hubbard, Oliver Payson (1809–1900). Dartmouth College, 1837–1866.

Oliver P. Hubbard was born in Connecticut but attended Hamilton College before graduating from Yale in 1828. After spending three years teaching in New York and Virginia, Hubbard returned to Yale as assistant in the laboratory of Benjamin Silliman (*q.v.*), one of whose daughters he married. While assisting in the laboratory, Hubbard also lectured at nearby Wesleyan University (1835). In 1837, he became professor of chemistry, mineralogy, and geology at Dartmouth College, holding that position until 1866, after which he served the school as lecturer (1866–1869) and professor of chemistry and pharmacy (1871–1883).

In 1837 Hubbard received an M.D. from the Medical College of Charlestown, South Carolina. His publications include a number of lectures on medical subjects and a history of Dartmouth Medical College. Hubbard was a member of numerous natural history societies, medical societies, the AAAS, and even the AHA. At various times, he served as secretary, vice-president, and president of the Academy of Sciences of New York, where he lived as professor emeritus from 1883 to 1900.

BDA.

Jackson, Isaac Wilber (1804–1877). Union College, 1831–1877.

Isaac W. Jackson, a Quaker, spent his entire adult life at Union College. After graduating from Union in 1826, he became a tutor from 1826 to 1831, and then professor of mathematics and natural philosophy from 1831 until his death in 1877. Although an enthusiastic floriculturist who raised rare species in his garden and corresponded with

English horticulturists especially about his 300 varieties of roses, Jackson did not publish or teach in natural history. His published works consist of four books—*Elementary Treatise on Mechanics* (Albany, 1852), *Elements of Trigonometry* (Schenectady, 1860), *Elements of Conic Sections* (Schenectady, 1857), and *Elementary Treatise on Optics* (New York, 1848). All four books grew out of his lectures at Union College and were intended for use by its students, although two of them became extremely popular elsewhere. *Elements of Conic Sections* went through eight editions; and the *Elementary Treatise on Optics,* which went through three editions, replaced Bache's edition of Brewster's *Optics* as the prescribed text at many American colleges and even one British university. A brilliant teacher, Jackson was called by Joseph Henry the "truest and most generous soul he ever knew."

BDA.
Potter, Eliphalet Nott. 1877. *Discourses Commemorative of Professor Taylor Lewis . . . and of Isaac W. Jackson* (Albany).
1877. *Semi-Centennial Anniversary of the Connection of Professor Isaac W. Jackson, LL.D. with the Faculty of Union College, 1876* (Albany).

Johnston, John (1806–1879). Wesleyan University, 1837–1873.

Born in Bristol, Maine, John Johnston later wrote a history of his hometown. He graduated in 1832 from Bowdoin College, where he insisted that his interest in science was kindled by Parker Cleaveland (*q.v.*). In his senior year, he became a member of the Methodist Episcopal Church and, upon graduation, taught at the Oneida Conference (Methodist) Seminary in Cazenovia, New York, where newly formed Wesleyan University found many of its first professors. In 1835 Johnston went to Wesleyan as associate professor of mathematics and in 1847 became professor of natural science, a position he held until 1873.

Johnston's early papers on mineralogy and physics appeared exclusively in the *AJS;* but he was best known for his widely used textbooks. His first, *A Manual of Chemistry on the Basis of Dr. Turner's Elements of Chemistry . . .* (Phila., 1842), was a careful and systematically arranged text, which was used in a number of colleges, including Bowdoin and Wesleyan. Dedicated to his teacher Parker Cleaveland, Johnston wrote, in this book, one of the country's earliest college texts to incorporate the elements of analytical chemistry. His other book, *Elements of Chemistry for the Use of Schools* (Phila., 1852), filled the void that had existed between the non-technical essays and the unwieldy comprehensive compendia on the subject. Both textbooks went through three editions.

An original member of the AAAS (1848), Johnston was elected to the APS in 1876.

Cleaveland, Nehemiah. 1882. *History of Bowdoin College with Biographical Sketches of its Graduates* edited by Alpheus S. Packard (Boston), pp. 432–433.

Joslin, Benjamin Franklin (1796–1861). Union College, 1822–1837.

After graduating from Union College in 1821, Benjamin F. Joslin remained as a tutor in mathematics. In 1826 he received a medical degree from the College of Physicians and Surgeons in New York City. Serving as professor of mathematics and natural philosophy at Union from 1827 to 1837, he then returned to New York City to hold the same position from 1838 to 1844 at New York University, where he was a quite unsuccessful teacher. The last two decades of his life may have been spent in private medical practice.

While most of Joslin's written work is in his speciality of homeopathic medicine, he apparently had extensive scientific interests. He edited the American edition of Dionysius Lardner's *A Treatise on Hydrostatics and Pneumatics* (Phila., Carey & Lea, 1832), published his own *Meteorological Observations and Essays* (Albany, Packard & Van Benthuysen, 1836), and wrote an article on "Irradiation," [*Trans. Amer. Philos. Soc.,* n.s., 4 (1831): pp.340–349.] In addition, the American Philosophical Society has drawings of Joslin's suggested improvements for an air pump [MSS., Oct. 25, 1823, and April 1, 1836]; and a letter from Stephen Alexander (*q.v.*) [April 16, 1831; Princeton University Library] indicates that Joslin assisted him in observing artificial star occultations. He joined the AAAS in 1849.

1859. *Trubner's Bibliographical Guide to American Literature* (London), p. 140 (a list of his medical pamphlets).
Jones, Theodore Frances, editor. 1933. *New York University, 1832:1932* (New York), *passim.*

Lasell, Edward (1809–1842). Williams College, 1833–1851.

Edward Lasell was born in Scholharie, New York, where his early studies were conducted under private tutors. After graduation as valedictorian of the class of 1828 at Williams College, Lasell taught high school in Pittsfield for a year until he became tutor at Williams where, in 1833, he was elected professor of chemistry. In 1851 Lasell resigned to establish at Auburndale, Massachusetts, a girls' school, known as the Lasell Female Seminary. He died the following year at age 43.

Cyclopaedia of American Biography **18**: p. 250.
Durfee, Calvin. 1871. *Williams Biographical Annals* (Boston), pp. 148–149.
1848. *Addresses at the Dedication of the New Cabinet and Observatory of Amherst College* (Amherst, Mass.), p. 45 (his views on education).

Lovering, Joseph (1813–1892). Harvard College, 1838–1888.

After his graduation from Harvard College in 1833, Joseph Lovering attended Harvard Divinity School, tutored in mathematics, and conducted prayers in the College Chapel. In 1838 he succeeded John Farrar as Hollis Professor of Natural Science, a position he held for a half-century. Active in the intellectual life of Boston, he delivered lectures at the Lyceum and the Lowell Institute on physics and astronomy and served as president of the American Academy of Arts and Sciences (1880–1892), in addition to editing six volumes of its *Memoirs*. As permanent secretary of the AAAS from 1854 to 1873, he edited fifteen volumes of its *Proceedings* and, in 1873, served as president of the Association.

Although most effective as editor rather than original investigator, Lovering published some sixty papers on astronomy and physics. Between 1867 and 1876 he had charge of determining longitudes in the United States by telegraphic communication with England under the auspices of the United States Survey, and his methods appear in "On the Determination of Transatlantic Longitudes by means of the Telegraphic Cables" (*Memoirs of the American Academy of Arts and Sciences*, n.s., **9** (1867): pp.437–477). Quite remarkably, in his sixty-second year, Lovering wrote a paper, "On a New Method of Measuring the Velocity of Electricity" (*Proc. AAAS*, **24** [1876]: pp. 35–39), which demonstrated his ability to keep abreast of the latest scientific developments; and though not a major contribution, the paper was even reprinted in the French *Journal de Physique* (Tome **6**). Lovering wrote no books; but his work at the Harvard Magnetic Observatory appeared as an entire volume of the *Memoirs of the American Academy of Arts and Sciences* (**10**), entitled "An account of the Magnetic Observations made at the Observatory of Harvard University . . ." (Cambridge, 1845). It is clear from this work that Lovering was aware of the problems in terrestrial magnetism; and he was able to interest Harvard students in the subject, many of whom assisted him at the Observatory, especially in the painstaking work of recording magnetic dips on the international "term" days.

While not as important as other Harvard scientists of the late nineteenth century, Lovering was an exceptionally determined worker and eloquent teacher. In 1873, he was elected to the NAS.

DAB.

Pierce, B.O. 1909. *NAS Biog. Mem.* **6**: pp. 327-344. (Contains a complete bibliography.)

McCulloh, Richard Sears (1818-1894). Princeton University, 1849-1854.

Richard S. McCulloh, son of the Baltimore banker in the McCulloch *vs.* Maryland case of 1819, went to school in Philadelphia, attended the University of Maryland, and graduated in 1836 from Princeton, where he was one of Joseph Henry's assistants. During the winter of 1838-1839, he studied chemistry in Philadelphia with James Curtis Booth, sometime lecturer at the University of Pennsylvania and the first German-trained analytical chemist in America. While in Philadelphia, McCulloh also met Alexander D. Bache (*q.v.*) and became an assistant observer at the Magnetic Observatory at Girard College. Undoubtedly, it was Bache who arranged all of McCulloh's later government appointments.

From 1841 to 1843, McCulloh was professor of science at Jefferson College, beginning his researches on the sugar industry. From 1846 until 1849 he served as melter and refiner of the United States Mint at Philadelphia, where he developed a new method for gold refining, the priority of which became a topic of some debate. In 1849 McCulloh became professor of natural philosophy at Princeton; and in 1854, he went to Columbia College, only to be expelled in 1863 when he returned to the South. During the Civil War, he worked as consulting chemist of the Confederate Nitre and Mining Bureau, where he improved the southern munitions manufacture. And after the conflict, he taught at Robert E. Lee's Washington College (later Washington and Lee) from 1866 to

1877, and then at Louisiana State University from 1877 to 1884.

McCulloh published one textbook, *Treatise on the Mechanical Theory of Heat and Its Applications to the Steam Engine,* which he dedicated to Joseph Henry (New York, D. Van Nostrand, 1876); and the list of his publications in applied science is considerable. Beginning in 1839, when he did a hydrodynamic survey of the Allegheny summit of the Chesapeake and Ohio Canal, McCulloh was involved in numerous projects. In 1843 he published a *Plan of Organization for the Naval Observatory* (Washington, 1843); and in 1846, he was elected to the APS, serving on its committee to review papers on "imponderable agents" and the committee to determine the solar parallax. His most important work was a number of reports on the sugar industry, prepared for the secretary of the treasury. The final combined version of these reports, 653 pages in length, was the nearest thing possible to a textbook on everything known about sugar. In it, besides recording information gained from travels to the South and to Cuba, McCulloh carefully evaluated all known European processes, including Biot's recent determination of sugar concentration by the rotation of plane polarized light.

Thomas, Milton H. 1947. "Professor McCulloh of Princeton, Columbia, and Points South." *Princeton University Library Chronicle* **9**: pp. 17–29. (Admirably fills the void created by the fact that McCulloh, a Southern sympathizer, has been systematically omitted from Northern biographical works.)

1848. *Reports From the Secretary of the Treasury of Scientific Investigations in relation to Sugar and Hydrometers . . .* (Washington).

Cajori, Florian. 1890. *Teaching and History of Mathematics in the United States* (Washington), p. 125. (Suggests that McCulloch (*sic*) revised Bartlett's *Mechanics,* although he did not allow his name to appear in the text.

Olmsted, Denison (1791–1859). Yale College, 1825–1859.

Denison Olmsted, son of an East Hartford farmer, prepared at home with the help of the local pastor to enter Yale College. After graduating in 1813, he taught in a New London boys' school and then returned to Yale in 1815 as tutor and student of theology under President Timothy Dwight. In 1817 Olmsted was appointed professor of chemistry in the University of North Carolina, where he also suggested the first state geological survey, became state geologist and mineralogist (1822), and eventually published the first reports of the survey (Raleigh, 1825).

In 1825 Olmsted returned to Yale as professor of mathematics and natural philosophy, although mathematics was dropped from his title in 1836, after which he taught principally astronomy until his death in 1859. Obvious from his 1816 master's oration on the need for normal schools in Connecticut was Olmsted's lifelong concern with public and popular education. He served on the Board of Commissioners on Education in Connecticut, and three of his textbooks were designed for the lower school grades and the general public: *A Compendium of Natural Philosophy* (New Haven, 1833), *A Compendium of Astronomy* (New Haven, 1839), and *Rudiments of Natural Philosophy and Astronomy* (New York, 1844). He also wrote two college textbooks, which grew out of his printed lecture outline for Yale students and his notes: *An Introduction to Natural Philosophy* (New Haven, 1831) and *An Introduction to Astronomy* (New York, 1839). All of his books were extremely popular, and his simpler books even found readers in Great Britain. However, his college texts were too hastily conceived, lacked the sophistication of the newer texts based on analytical rather than purely geometrical methods of demonstration, and were made more suitable for use in college courses in the later revisions by Ebenezer Snell (*q.v.*).

Olmsted published more than fifty papers, over thirty appearing in the *AJS*, primarily on chemistry and meteorology but also in geology and mineralogy. His most celebrated scientific work, however, was in astronomy, and consists of his explanation of the extra-terrestrial or celestial nature of meteor showers ("Observations on the Meteors of November 13th, 1833," *AJS*, **25** (1834): pp. 363–411; continued in **26** (1839): pp. 132–174) and the discernment of a new periodicity in the aurora borealis ("On the Recent Secular Period of the Aurora Borealis," *Smithsonian Contributions to Knowledge*, **8** [1856], 52 pp.).

Olmsted was a founding member of the AAAS (1848).

DAB.

Daniels, George H. 1968. *American Science in the Age of Jackson* (New York), pp. 219–220.

1859. "Address of Professor Alexis Caswell, President." *AAAS, Proc.* **13:** pp. 4–7.

Dexter, Franklin Bowditch. 1912. *Biographical Sketches of the Graduates of Yale College* (New Haven) **6:** pp. 592–600. (Includes complete bibliography.)

Park, Roswell (1807–1869). University of Pennsylvania, 1836–1842.

Born in Connecticut, Roswell Park moved to New York, where he attended Hamilton Academy and Hamilton College. He received an A.B. from Union College and graduated at the head of the class of 1831 at the United States Military Academy. For five years, Park served in the military in the construction of forts and dams. In 1836 he resigned to become professor of chemistry and natural philosophy at the University of Pennsylvania in between A.D. Bache's (*q.v.*) two appointments.

While at the University of Pennsylvania, Park published the first of his two volumes of poetry, a history of the United States Military Academy, and *An Outline of Magnetism* (no title page [1840], University of Pennsylvania Library) in which he appended twenty pages of questions to be answered by students in conjunction with their textbooks. He also wrote a *Pantology; or a Systematic Survey of Human Knowledge* (Phila., 1841), a curiously late attempt to define and relate all branches of human knowledge in one volume. Although this book remains valuable for its extensive bibliography, there is no record of who may have bought it, although, by 1847, it was in its fourth edition.

After Park resigned from the University of Pennsylvania in 1842, he entered the ministry of the Protestant Episcopal Church. In addition to these duties, he spent most of the remainder of his life as an educator in preparatory schools and colleges in the Midwest. His association with all of these schools was distinctly marked by a devotion to the inclusion of scientific studies and to provision for special scientific electives and programs. The father of the famed surgeon of the same name, Roswell Park was a member of the APS and an original member of the AAAS (1848).

DAB.

Parker, William B. (1809?–1889). Middlebury College, 1848–1881.

William B. Parker graduated from Middlebury College in 1830. He spent the next eighteen years teaching in various academies, tutoring at Middlebury (1832-1834), and studying at Andover Theological Seminary (1834–1836). In 1848,

Parker was appointed professor of mathematics and natural philosophy at Middlebury, where he taught until his retirement in 1881. Parker seems to have left no scientific works and was not even a member of the AAAS.

1890. *Catalogue of the Officers and Alumni of Middlebury College* . . . (Middlebury, Vt.), p. 67.

Sabin, in the *Bibliotheca Americana* (14: p.197) indicates that there was a William B. Parker attached to Capt. Marcy's Texas Exploring Expedition (1854), but I cannot locate the notes to ascertain whether this is the same Parker.

Pearson, Jonathan (1813–1887). Union College, 1850–1887.

Jonathan Pearson spent his entire adult life at Union College. He graduated there in 1835; became a tutor from 1836–1839; adjunct professor of chemistry and natural history, 1839–1850; full professor, 1850–1857; professor of natural history, 1857–1873; and professor of agriculture and botany, 1873–1887. When he was first asked to teach chemistry in 1838, Pearson wrote in his famous diary that "Chemistry [is] a study which I never before studied"; but he learned the subject on his own, as it was then possible to do, and became a successful teacher (June 8 and 9, 1838). In 1846 Union President Nott called Pearson to the attention of Joseph Henry when the latter was looking for people to fill posts at the Smithsonian Institution, but Pearson never received an appointment there.

The author of five books on Albany, Schenectady, the genealogy of their settlers, and the local Dutch Reformed Church, Pearson joined the AAAS in 1849.

BDA.
Pearson MSS Diary, Union College.

Pierce, Benjamin (1809–1880). Harvard College, 1833–1880.

Benjamin Peirce was the most celebrated American mathematician of the nineteenth century. The list of his achievements in astronomy, analytical mechanics, number theory, and especially linear associative algebra, would strain the purposes of these biographies.

Peirce was a member of one of New England's oldest families. Born in 1809 in Salem, he attended the Private Grammar School there, meeting Nathaniel Bowditch, whom he later aided in a translation and revision of LaPlace's *Mécanique Céleste*. Bowditch, then the most famous American man of mathematics, quickly recognized Peirce's genius.

It appears that from an early age Peirce had only the strongest determination to become a mathematician. After graduating from Harvard in 1829, he taught mathematics at a private school and served as tutor at Harvard until, in 1838, he was named professor of astronomy and mathematics. In 1842 Peirce became Perkins Professor of Mathematics and Astronomy at Harvard, remaining there until his death in 1880 after an association with the school of almost half a century.

Peirce and Harvard were both unique phenomena in the 1830's. Peirce had none of the teaching ability or patience of the typical college teacher of the period. He was interested in brilliant mathematicians alone, and not in the general student. By 1838, he had convinced the Corporation that mathematics should become an elective after the elements were learned in the freshman year. Immediately, enrollment in higher courses decreased; Peirce could teach only the higher courses decreased; Peirce could teach only the interested; and the uniquely large Harvard teaching staff could instruct the remainder—a situation much like that which prevailed in the later University.

From 1835 to 1855, Peirce wrote a number of textbooks on mathematics, sound, and analytic mechanics. Most never became popular, since they showed no concern for student needs. They did, however, introduce certain novelties, such as the use of infinitesimals and limits in elementary books, which were quickly incorporated by more successful authors.

Besides providing the only instruction in America in the most sophisticated contemporary mathematics, Peirce had more profound effects upon Harvard. He was a potent force in the founding of the Harvard Observatory (1843) and in the operation of the Lawrence Scientific School (1846). Often known as the "Father of American Mathematics," Peirce established and edited the *Cambridge Miscellany of Mathematics, Physics, and Astronomy* (1842) and was adviser to the *American Nautical Almanac,* established in Cambridge probably to be near him (1849). He was a member of the United States Coast Survey (1852–1874) and its superintendent from 1867 to 1874. Peirce was honored by membership in many European societies, including the Royal Astronomical Society (1850), for his work on the gravitational perturbations of the planets. A member of all major American societies, he was a founding member of the NAS (1863), where he presented the papers destined to become his most famous book, *Linear Associative Algebra* (1870).

DAB is the most obvious starting point for the mass of Peirce material.
Cajori, Florian. 1890. *The Teaching and History of Mathematics in the United States* (Washington), pp. 133–147.

Renwick, James (1792–1863). Columbia College, 1820–1853.

Born in Liverpool, James Renwick came to America with his family when he was only a few years old. After graduating from Columbia College at the head of the class of 1807, he continued his education in Europe, traveling with his long time friend Washington Irving. Although Renwick first delivered lectures on natural philosophy at Columbia College in 1814, it was not until 1820 that he was appointed professor of natural philosophy and experimental chemistry, a position he held until he became Columbia's first emeritus professor in 1853. Before joining Columbia permanently, Renwick had gained valuable experience as a topographical engineer in the government service and as engineer in the state militia.

Within a few years, Renwick became the most famous engineer in America. Rarely was any project of importance, public or private, undertaken without consulting him. His special areas of interest were canal construction, steam-boiler operations, and, later, railroad planning. The Franklin Institute in Philadelphia awarded him a medal in 1826 for his novel use of cradles and inclined planes in the Morris Canal, which connected the Delaware and Hudson rivers; in 1838 he served on a presidential commission to find ways to make steam boilers safe from explosion; and in 1840, he worked on a three-man commission to draw the Maine boundary.

Renwick was one of the most prolific authors of his generation. His first textbook, *Outlines of Natural Philosophy* (2 v., New York, 1822–1823) consisted of his lectures at Columbia and, though neither original nor particularly well done, represents the first full-length text on this subject by an American author. His *Elements of Mechanics* (Phila., 1832), again formed essentially from his lectures at Columbia, is considerably more mathematical and contains a unique chapter on canals. Renwick was responsible for two works on the steam engine; a popular revision of Lardner's *Steam Engine*

(Phila., 1828 & 1836); and his own highly technical and original *Treatise on the Steam Engine* (New York, 1830), which was even translated into German and French in 1842.

In 1840 Renwick published *First Principles of Chemistry* (New York, 1840) as a text for college students, making it considerably simpler than larger treatises, since the subject was then "included among the studies of the lower classes, instead of being contained, as formerly, to the higher only." He also edited the American editions of Parke's *Rudiments of Chemistry* (1824) and Daniell's *Chemical Philosophy* (1840).

The remainder of Renwick's vast bibliography consists of treatises on artillery and the practical applications of mechanics. Upon his retirement in 1853, he selected the exhibits for the New York Industrial Exhibition of 1853 and wrote a number of biographies, including lives of David Rittenhouse, Robert Fulton, Benjamin Thompson, John Jay, DeWitt Clinton, and Alexander Hamilton. Renwick was one of the most remarkable men of his time, equally at home among students, industrialists, and scientists, and capable of writing both simple essays and serious technical works.

DAB.

Rogers, Henry D. (1808–1866). University of Pennsylvania, 1835–1846.

Henry D. Rogers was born in Philadelphia and educated both by his father and by schools in Maryland and Virginia. His first job was as chemistry lecturer in Baltimore's Maryland Institute in 1828. Appointed professor of chemistry and natural philosophy at Dickinson College in 1830, he founded and edited there a monthly journal entitled *Messenger of Useful Knowledge* (1830–1831). After one year at Dickinson, Rogers became interested in socialism and accompanied Robert Dale Owen to London, where he became interested in geology, associated with De la Beche and Lyell at the Geological Society, and attended the chemical laboratory of Sir Edward Turner.

On his return in 1833, Rogers gave geology lectures at the Franklin Institute in Philadelphia and worked on a New England railroad survey, In 1835 he was appointed professor of geology and mineralogy at the University of Pennsylvania. In the same year, Rogers was elected to both the APS and the Academy of Natural Sciences in Philadelphia, besides being appointed by the legislatures of Pennsylvania and New Jersey to make geological surveys of their states.

While at the University of Pennsylvania, Rogers printed *A Guide to a Course of Lectures on Geology Delivered in the University of Pennsylvania* for the use of students. Although he soon left the New Jersey Survey, he remained with the Pennsylvania Survey until it was disbanded in 1842. He helped organize in 1840 the Association of American Geologists, the forerunner of the AAAS.

After leaving the University of Pennsylvania, Rogers traveled to Boston to work on the final report of the Pennsylvania Survey, which G.P. Merrill called the "most important document on the geology of America that had appeared by that time (1868)." While in Boston, Rogers revealed himself as an early champion of "developmentalism"; and consequently, his appointment to the Harvard faculty was blocked by Asa Gray, who disapproved of Rogers's advocacy of that theory. In 1855 he moved to the University of Glasgow to find better publication facilities for his work. There he was appointed Reguis Professor of Natural History and became president of the Philosophical Society of Glasgow, where he remained until his death in 1866.

Beginning in 1834, Rogers published dozens of papers in the *APS, Trans.*, the *AJS,* and the *British Association Reports,* among other journals. That Rogers enjoyed a wide reputation is indicated by his membership in the Royal Society of Edinburgh and the Geological Society of London. His international fame was secured by an oral report he gave at the Association of American Geologists in 1842 at its Boston meeting, which was attended by Sir Charles Lyell. Here, as a result of his studies on the Appalachian Range, Rogers advanced the wave theory of mountain chains—the first suggestion that all mountain chains are characterized by the same features. While incorrect in some details, it marked the first important American contribution to dynamical geology.

DAB.

Dupree, A. Hunter. 1954. *Asa Gray* (Cambridge, Mass.), pp. 144, 149.

Ruschenberger, W.W. 1885. *A Sketch of the Life of Robert E. Rogers. . . with Biographical Notices of his Father and Brothers* (Philadelphia), pp.18-23 and *passim*. Bibliography on pp. 41-43.

Shepard, Charles U. (1804–1886). Amherst College, 1847–1877.

A native of Rhode Island, Charles U. Shepard began his college career at Brown, but transferred to Amherst in 1820 to become a member of its original student body. At that time, Amos Eaton was lecturing at Amherst; and he utilized in his lectures many minerals which Shepard had collected as a student. After graduating from Amherst in 1824, Shepard taught natural science in Boston schools and studied under Thomas Nuttall, the curator of the Harvard Botanic Museum. At this time, he began to collect minerals in Maine and submitted the first of some forty articles to *Silliman's Journal.*

In 1827 Shepard moved to New Haven to become one of the assistants in Silliman's laboratory, and in 1830 he delivered lectures in natural history at Yale, a task he undertook sporadically until 1844. From 1830 to 1845 Shepard traveled extensively in this country and abroad on various scientific missions. With Silliman's recommendation, he was chosen by the secretary of the treasury to survey the sugar refining industry in the southern states; and his report, *Manual on the Cultivation of the Sugar Cane, and the Fabrication and Refinement of Sugar* (Washington, printed by Francis P. Blair, 1833), which inaugurated his continuing interest in scientific agriculture, is celebrated as the first example of federal government sponsored analytic research, This report and his *Report on the Geological Survey of Connecticut* (New Haven, 1837) secured him a widespread reputation.

In 1839 Shepard began a series of trips to Europe, which, by his exchange of minerals and knowledge of American mineralogy, secured him election to numerous foreign societies as well as perhaps the finest personal cabinet of minerals in the United States. In 1847 he severed his loose connections with midwestern geological surveys, the medical College of Charlestown, South Carolina, and Yale, to become full-time Massachusetts Professor of Chemistry and Natural History at Amherst College, a position he held until 1877. Shepard's most celebrated work was a *Treatise on Mineralogy* (New Haven, Hezekiah Howe, 1832; 2nd part, 1835), which was revised in 1839 and again in 1857.

DAB.

Daniels, George H. 1968. *American Science in the Age of Jackson* (New York), pp.222–223 and *passim.*

Silliman, Benjamin (1779–1864). Yale College, 1802–1853.

A native of Connecticut, Benjamin Silliman graduated from Yale College in 1796, taught in private school for two years, and returned to Yale as tutor in 1799. At the same time, he studied law and was admitted to the bar in 1802. Meanwhile, Yale had invited this bright young man to become proficient in chemistry and natural history, areas which had been sadly neglected there. Since only the medical schools of Harvard and the University of Pennsylvania offered competent instruction in chemistry, Silliman traveled to Philadelphia to attend the lectures of Woodhouse, Rush, Barton, Wistar, and Shippen on chemistry, anatomy, physiology, and other medical subjects. While at the University of Pennsylvania, he made a lifelong friend in Robert Hare, also a student, whom Silliman assisted in his early researches with his world-renowned blowpipe. From Philadelphia, he made excursions to Princeton College to converse with John Maclean, the recently imported devotee of the new chemistry.

Silliman returned to Yale to deliver his first lectures as professor of chemistry and natural history in 1804, but soon left to continue his education in London and Edinburgh. On his return, he wrote a book about his European travels and began a career with so many associations that it requires books to unravel them. In 1806 he published a paper on the geology of New Haven—the first of some sixty scientific papers on geology, mineralogy, chemistry, and natural history. In 1811 he subjected previously unexplored minerals to Hare's blowpipe analysis and made novel observations of the voltaic current in a chemical deflagrator, his most original scientific work.

In 1818 Silliman planned the *American Journal of Science,* which first appeared in 1819 and which he edited for many years. Originally conceived upon certain European models as an attempt to combine interest in physical sciences and natural history, the *Journal* quickly took on the peculiar character of American needs and interests.For many years, most of the articles were on natural history, mineralogy, and geology. Silliman's own editorial bias gave it a distinctly "utilitarian" flavor, but it succeeded in rousing public awareness of science and invigorating communication between scientists. The speed with which its manuscripts appeared in print would shame many modern editors.

Through the *Journal* Silliman had correspondence with virtually every scientist of the first half of the nineteenth century, and it is a rare old college library that does not contain some examples. Not only did thousands of Yale students pass through his laboratory, but so too did others seeking a course of chemical lectures. The laboratory received specimens from all over New England and from the exploring expeditions to analyze. In addition to his lectures at Yale, Silliman gave countless popular lectures, especially from 1834 to 1845, in cities as far away as New Orleans and St. Louis, interesting students in science across half the country.

Silliman was a sincerely religious man, somewhat occupied, especially in his earlier years, with increasingly complex problems of science and religion. Because of his prominence in the life of science, he was often cited as one of the chief promoters of cooperations between the two disciplines, although in later life he was content to let others wage any active battles. By midcentury, his view "that there is no proof of the truth of that [Biblical] history other than that afforded by geology"(A.L.S. to Oliver P. Hubbard [*q.v.*], Jan. 21, 1856, Yale University Library) was not upheld in that form by the scientific community.

Silliman, quite naturally, belonged to every well-known society in America, to European societies, and to the group of fifty original members of the NAS (1863). Though not a brilliant scientist, Silliman was nevertheless a thorough investigator and the most influential member of the scientific community of his day.

Caswell, Alexis. 1877. *NAS Biog. Mem.* **1:** pp. 99–112.

See *DAB* and Daniels, *op. cit.,* p.223 for some indication of the vast amount of printed material and manuscript sources relating to one of the most important figures in American science.

Silliman, Benjamin, Jr. (1816–1885). Yale College, 1846–1870.

Son of the famous B. Silliman the Elder (*q.v.*), Benjamin Silliman, Jr., also attended Yale, where he studied with his father. After his graduation in 1837, he became an assistant in the elder's famous laboratory. The following year, he joined the Staff of the *AJS*, remaining an editor for the *Journal* until his death in 1885. In 1842 he provided instruction to his father's advanced students on a private basis; and this self-supporting operation has been recently described as the origin of the arrangement to which the Yale trustees subscribed in 1846 as a prelude to the Sheffield Scientific School.

In 1846, just after he had become professor of practical chemistry in the College, Silliman initiated the proposal, which his father presented to the trustees, that a chair of agricultural chemistry be established. Silliman had lectured in New Orleans in agricultural chemistry during the winter of 1845-1846 and John Pitkin Norton, who was the first to occupy Yale's chair, studied in his private laboratory.

From 1849 to 1854, Silliman served as professor of chemistry in the medical department of the University of Louisville, a position he also held during his associations with Yale. He was especially active outside of college as a consulting chemist on analytical and mining problems. His Report on the *Rock Oil, or Petroleum, from Venango County, Pennsylvania* (1855), the site where America's first well was drilled, proved a masterpiece of industrial chemistry. In it, Silliman shows how the various factions can be separated and employed for a variety of purposes—the beginning of a sophisticated petrochemical industry. Unfortunately, a dispute with some scientists in southern California who had failed to locate oil in an area Silliman had suggested, resulted in some problems. Silliman was embarrassed into resigning from Yale College in 1870; and was also removed from the Governing Board of Yale's Sheffield Scientific School, although he was associated with its medical department until his death in 1885.

Silliman belonged to the American Academy of Arts and Sciences and to the AAAS. He was also a founding member of the NAS (1863), although his enemies tried to force his resignation after the "oil affair." The author of more than fifty papers on chemistry and mineralogy generally printed in the *AJS,* Silliman also wrote two college textbooks: *First Principles of Chemistry* (1847) and *First Principles of Physics* (1859). Fortunately, within his lifetime Silliman was vindicated from the disgrace of the "oil affair" when an oil industry finally developed in southern California.

DAB.
White, Gerald T. 1968. "Benjamin Silliman, Jr. and the Origins of the Sheffield Scientific School." *Ventures* 8: pp. 19-25. (Suggests that earlier biographical accounts are somewhat distorted because of the influence of Silliman's enemies.

Smith, Augustus William (1802-1866). Wesleyan University, 1831-1857.

Augustus W. Smith graduated from Hamilton College in 1825 and taught in the Oneida Conference (Methodist) Seminary at Cazenovia until he was appointed first professor of mathematics and astronomy at Wesleyan University in 1831. He remained at Wesleyan until 1859, when he became professor of natural philosophy at the United States Naval Academy, teaching there until his death in 1866.

Smith received a commission on the government expedition to Labrador to observe the solar eclipse of 1862. His only textbook, essentially completed by 1849, was not published until 1857, when it appeared as *An Elementary Treatise on Mechanics Embracing the Theory of Statics and Dynamics, and its applications to solids and fluids, prepared for the undergraduate course in Wesleyan University* (New York, 1857). The text makes free use of the calculus and is one of the first in America to follow the newer analytical arrangement.

BDA.

Smyth, William (1797-1868). Bowdoin College, 1828-1868.

The son of a poor shipbuilder in Pittston, Maine, William Smyth prepared for college on his own and graduated from Bowdoin with honors in the class of 1822. After graduation, he spent a year studying theology at Andover Seminary before returning to Bowdoin as Greek instructor in 1823. The following year, Smyth became tutor of mathematics and began his most successful teaching career as the first instructor in an American college to utilize the blackboard for classroom instruction. By 1828, Smyth was named professor of mathematics, a position he held until his death in 1868, although he also assisted in astronomy.

Smyth was the most successful mathematics textbook writer of his generation, originally preparing all of his texts for use by Bowdoin students. The first, a small book on trigonometry, was quickly followed by *Algebra* (Brunswick, 1830), *Elements of Plane Trigonometry* (Boston, 1834), and *Elements of Differential and Integral Calculus* (Portland, 1854). Smyth's first algebra was split into texts—one for schools, *Elements of Algebra* (Portland, 18??), and the other for colleges, *Treatise on Algebra* (Portland, 1852). The enlargement of his trigonometry with its application to surveying and navigation received the praise of A.D. Bache (*q.v.*) for its originality.

Smyth's books covered all topics taught in the American college—trigonometry, geometry, algebra, analytical geometry, and calculus; and all his works went through numerous printings, editions, and revisions. Some colleges used no other mathematics books than Smyth's, while many used at least one of them. The *Treatise on Algebra* was by far the most widely used algebra text in mid-nineteenth-century America.

A religious leader in his community, Smyth was also deeply involved in local problems. He was especially interested in popular education; and his home served as a station on the "underground railroad."

DAB.

Cajori, Florian. 1890. *The Teaching and History of Mathematics in the United States* (Washington).

Packard, Alpheus S. 1868. *Address on the Life and Character of William Smyth* (Brunswick, Maine).

Snell, Ebenezer (1801–1876). Amherst College, 1833–1876.

Ebenezer Strong Snell was a brilliant and methodical craftsman and teacher who spent his entire adult life at Amherst College. After his graduation in 1822, he taught in Amherst Academy until the College was empowered to grant the Bachelor of Arts degree, which he received in 1825. From then until 1853, he served as a tutor at Amherst College in mathematics and language and in 1833 became professor of mathematics and natural philosophy.

Although not a scientist of wide fame, Snell was extremely competent in his field. He drew countless detailed geometrical diagrams for the use of his students, many of which illustrated modern French analysis, and constructed much of the apparatus at Amherst. He published two articles on vibratory motion as applied to a waterfall and to the analysis of polarized light in the *AJS* (**28** [1859]: pp.228-231 and **32** [1861]: pp. 376-379), in addition to an article on the gyroscope. Amherst College Library has fairly extensive manuscript remains of Snell's, including calculations on a lunar eclipse of July, 1848, and a "Mathematical Diary (1822-60)" in which he collected geometrical problems of considerable complexity, interesting problems from the differential calculus, and his own solutions to the prize problems in the *Mathematical Monthly*. The extreme care with which Snell prepared his lectures, his diagrams, and his apparatus caused him to eliminate the redundancies in Denison Olmsted's *(q.v.)* textbooks and render them far more suitable to college students. He revised and prepared diagrams for *An Introduction to Natural Philosophy* (New York, Collins & Bros., 1863), and *A Compendium of Natural Philosophy, adapted to Use of Schools and Academies* (New York, Clark & Maynard, 1868).

1877. *Addresses at the Annual Meeting of the Amherst College Alumni Commemorative of the late Professor Snell* [June 27, 1877] (Springfield, Mass.).
Manuscripts in Amherst College Library.

Stanley, Anthony Drumond (1810-1853). Yale College, 1836–1853.

Anthony D. Stanley lived his entire life in Connecticut. After graduating from Yale College in 1830, he became tutor in mathematics at Yale in 1832 and professor of mathematics in 1836, holding that position until his death in 1853. Stanley helped President Jeremiah Day of Yale prepare the last edition of his *Algebra* (1852), although by this time Day's work was being used at few of the better colleges. He also published a textbook on his own, *An Elementary Treatise of Spherical Geometry and Trigonometry* (New Haven, 1848), which had limited popularity. Stanley's most useful work to students, who were now required to perform complicated calculations in physics and especially astronomy, was a compilation of seven-place *Tables of Logarithms of Numbers and of Logarithmic Sines* . . . (New Haven, 1841), reprinted in 1847, 1849, and 1851. Stanley left several works in manuscript when he died at the early age of forty-three in 1853.

ACAB.
Cajori, Florian. 1890. *The Teaching and History of Mathematics in the United States* (Washington), p. 63.

Tatlock, John (1808-1886). Williams College, 1838–1868.

Born in North Wales, John Tatlock grew up in Liverpool, where he worked in the office of a wine merchant. In 1830 he migrated to Philadelphia and then to New York, where he studied privately for college. After graduating valedictorian of the Williams College class of 1836, he remained at Williams as tutor and in 1838 was appointed professor of mathematics. He resigned in 1867 to join the bar and to conduct a law practice in Lenox, Massachusetts, until his death in 1886, leaving no record of scientific achievement.

Durfee, Calvin. 1871. *Williams College Biographical Annals* (Boston), p. 151.

Thompson, Zadock (1796–1856). University of Vermont, 1851–1856.

Son of a poor Vermont farmer, Zadock Thompson graduated from the University of Vermont in 1823 at the age of twenty-seven to remain in Burlington as schoolteacher and author. Appointed tutor at the University of Vermont in 1825, he soon displayed an interest in natural history, participated in the University's "College of Natural History" (1826), and edited a short-lived magazine in which he published his own conservative view of the conciliation of science with the Deluge. His school text, *Youth's Assistant in Practical Arithmetic* (Burlington, 1826), ran through fourteen editions.

During some two decades of travel, teaching, writing, and even performing as deacon in the Protestant Episcopal Church, Thompson kept alive his interest in natural history, building up a private cabinet of over three thousand specimens, later purchased by the University. He assisted Charles B. Adams (*q.v.*) on the state geological survey of Vermont especially in the measurement of mountain heights, becoming assistant state geologist in 1845 and state naturalist in 1853. Although best known for his general history of Vermont, which contains much natural history, Thompson also published a separate *Natural History of Vermont* (Burlington, 1850). In 1851 he returned to the University as professor of chemistry and natural history, holding that position until his death in 1856.

Adams, Charles Baker. 1845. *First Annual Report on Geology of the State of Vermont* (Burlington).
DAB.
Lindsay, Julian Ira. 1954. *Tradition Looks Forward* (Burlington).

Torrey, John (1796–1873). Princeton University, 1830–1854.

Descended from an old American family, John Torrey was fifteen when his father was appointed fiscal agent at Greenwich (Village) Prison. Amos Eaton, soon to become the first great popularizer of science in America, was then serving time in the prison for debt; and he gave Torrey the boy's first lessons in plant identification and dissection. By 1815, Torrey was studying medicine at the College of Physicians and Surgeons, where he earned his M.D. in 1818. At medical school, Torrey met Dr. Samuel L. Mitchill and others who together formed the Lyceum of Natural History (1817), the forerunner of the New York Academy of Sciences.

After medical school, Torrey entered private practice. He also prepared a catalog of plants growing near New York City (1817) and was soon corresponding with the leading botanists of Europe and America. Shortly after, his first two papers appeared in the initial volume of the *AJS*. The sucess of Torrey's botanical work led him to be recognized as unofficial cataloger of the plants of North America. All of the plants collected in the great western expeditions, beginning with Long's Expedition in 1820 and including Fremont's Expedition of 1842 and the Mexican Boundary Survey of 1848-1854, were shipped to Torrey, who spent his life carefully naming thousands of new species.

Although his fame was as a botanist, Torrey made his living as a physician, then as professor of chemistry, mineralogy, and geology at the United States Military Academy (1824-1827), and finally as professor of chemistry in the College of Physicians and Surgeons (1827-1855), simulataneously serving as professor of natural history at Princeton College (1830-1854). After leaving the academic world, Torrey spent the remainder of his life (1854-1873) as chief assayer in the United States Assay Office in New York—work which afforded him opportunity for extensive travel, especially to see the live flora of the West whose dried specimens he had inspected.

In 1836 Torrey was appointed New York state botanist; and his *Flora of the State of New York* (1843) became the most complete state botanical survey in the first seventy-five years of the century. He published other volumes and numerous papers on the flora of the United States, although he never combined the separate works into a comprehensive survey of North American plants. None of his works were intended as college texts.

DAB.
Gray, Asa. 1877. *NAS Biog. Mem.* 1: pp. 165—176.
Reingold, Nathan, editor. 1964. *Science in Nineteenth Century America: A Documentary History* (New York), pp. 42, 48, and *passim*.

Turner, Edward (1797–1838). Middlebury College, 1828–1838.

Born in North Haven, Connecticut, Edward Turner graduated from Yale College in 1818. After spending three years as teacher at Weston Academy and rector of Hopkins Grammar School, Turner studied for two years at Andover Theological Seminary (1821–1823). In 1823 he became tutor in Middlebury College, and in 1825 was named Painter Professor of Mathematics and Natural Philosophy at the College, holding that position until his early death in 1838. A bibliographical search reveals that Turner has left no scientific or literary remains.

Dexter, Franklin Bowditch. 1913. *Biographical Notices of Graduates of Yale College* (New Haven), p. 42.

Twining, Alexander Catlin (1801–1884). Middlebury College, 1839–1849.

Alexander C. Twining, son of the treasurer of Yale College, graduated from Yale in 1820. He spent the next few years in residence for his master's degree and attended the Andover Seminary, although he did not complete the theological course. He returned to Yale in 1823 to attend more lectures in natural philosophy and then went to the Military Academy for courses in civil engineering, mathematics, and astronomy.

While at West Point, Twining observed the meteor shower of November 13, 1833. He painstakingly collected reports on this shower from around the country; and, by tracing the trajectories of the meteoric pieces, he was able to conclude, independently of Denison Olmsted (*q.v.*), that the meteors were not affected by the earth's rotation and hence were a purely celestial rather than terrestrial problem. His results were printed in the *AJS* (**26**, 2 [1834]: pp. 320–352), where Olmsted reached the same result by consideration of the motion of the single radiant point over a limited area (*ibid.*, pp.132–174).

Between 1834 and 1839, Twining worked as a railroad surveying engineer, chiefly for the northern routes of the Hartford and New Haven Railroad Company. He published a few surveys and was for many years consultant to New England and also midwestern railways. His other publications are concerned with mathematical problems.

For the ten years from 1839 to 1849, Twining was professor of mathematics and natural philosophy at Middlebury College. After his resignation, he returned to industry, securing a number of patents including one for an ice manufacturing process. Twining joined the AAAS in 1865.

DAB.

Vethake, Henry (1792—1866). University of Pennsylvania, 1836-1855.

Henry Vethake was born in British Guiana but raised in this country, where he graduated from Columbia College in 1808. He then studied law and served as a tutor at Columbia until he became professor of mathematics and natural philosophy at Queen's College (Rutgers). He spent the next twenty years at various colleges—Princeton (1817-1820), Dickinson (1821-1824), Princeton again (1829-1832), University of the City of New York (1832-1833), and Washington College, Virginia (1835-1836). Finally he settled at the University of Pennsylvania, where he remained until 1859 when he moved to Philadelphia's Polytechnic College, teaching there until his death in 1866.

Vethake, perhaps uniquely, was not continuously professor of both mathematics and natural philosophy at the University of Pennsylvania, but professor of mathematics and philosophy from 1836 to 1855 and of philosophy alone from 1855 to 1859. Although he was a competent teacher of mathematics, his books and numerous contributions to periodicals are on political economy. In this field, he was the American pioneer, being the first to expound orthodox *laissez-faire* economics. After his *Introductory Lecture on Political Economy* (Phila., 1831), Vethake was elected to the APS; and his *Principles of Political Economy* (Phila., 1838) was the "economics" text in America for years.

DAB.

Webster, John White (1793-1850). Harvard College, 1827-1849.

John W. Webster, a wealthy Bostonian, graduated from Harvard College in 1811 and from Harvard Medical School in 1815. He then spent a number of years studying medicine in London and traveling. On his return, Webster wrote his first book, *A description of the Island of St. Michael, comprising an account of its geological structure* . . . (Boston, 1821), and finally was forced to find a job after his high living had exhausted his inheritance. In 1824 he was named lecturer in chemistry, mineralogy, and geology at Harvard and finally, in 1827, Erving Professor of Chemistry and Mineralogy, although he was required to teach in the medical school in addition to his undergraduate duties.

Webster has become famous for his conviction and subsequent execution for the macabre murder of Dr. George Parkman (uncle of historian, Francis), which became one of the famous cases in American legal history. Webster, it is true, was neither a brilliant chemist nor a particularly interesting teacher, but his important work has been glossed over because of the circumstances of his crime. His *Manual of Chemistry* (Boston, 1826; 2nd ed., 1828) was a compilation of the work of European authors arranged according to his lectures at Harvard; and it was adopted at Amherst and Brown among other colleges. Its third edition (Boston, 1839) had a considerably different format from earlier textbooks in America; the standard chapters on electricity and magnetism, now thought proper to the natural philosophy department alone, were omitted and in their place were included extended sections on animal chemistry and organic analysis. A few years later, Webster edited the American edition of the most influential chemistry text of the era, Justus Liebig's *Organic Chemistry in its Applications to Agriculture and Physiology* . . . (Cambridge John Owen, 1841). The notes suplied to this work show Webster to be familiar with American experiments in scientific agriculture.

Webster arranged the 26,000 specimens in Harvard's mineral collection (1827–1836), was associate editor of the Boston Journal of Philosophy and the Arts (1824–1826), and an original member of the AAAS (1848).

DAB.

Forster, John. 1874. *Life of Charles Dickens* (4 v., Philadelphia) 3: pp. 402–403.

Morison, Samuel E. 1937. *Three Centuries of Harvard, 1636–1936* (Cambridge, Mass), pp. 282–285.

Quincy, Josiah. 1840. *History of Harvard University* (2 v., Cambridge, Mass.) 2: pp. 383, 401.

For the vast amount of material relating to Webster's crime and trial, see *DAB* for bibliography, Dicken's comments for literary example of the contemporary excitement, and Morison for condensed historical account.

Wyman, Jeffries (1814–1874). Harvard College, 1847–1874.

Son of a prosperous Massachusetts physician, Jeffries Wyman graduated from Harvard College in 1833 and from the Medical College of Boston in 1837. He immediately joined the Boston Society of Natural History, where he served as officer or curator for most of his life. From 1840 to 1841 he was curator and lecturer on anatomy at the Lowell Institute until he traveled to Paris to study with the French masters. On his return, he was called to fill the chair of anatomy and physiology in the medical department of Hampden-Sydney College. Four years later in 1847, he returned to Harvard to spend the remainder of his life as Hersey Professor of Anatomy.

Wyman's appointment at Harvard was not in the Medical School as had been the earlier Hersey appointments. Instead, the chair of anatomy and surgery was split, and Wyman's only teaching responsibilities were in the college in Cambridge. Here he taught comparative anatomy and physiology to the entire senior class and gave advanced instruction to special students. His prepared diagrams and free-hand lecture drawings were especially applauded by students.

Wyman's early publications, from 1839 to 1844, were in physiology, paleontology, and zoology. Later, he contributed to the fields of ethnology and archaeology, and by the time of his death, was the leading anthropologist in America, traveling to the American South, South America, and Labrador to collect animals. His studies on symmetry and homologous parts in the animal kingdom were especially noteworthy; and although Wyman chose not to participate in the raging Darwinian battles, he upheld the evolutionary hypothesis and gave it support through his studies on vestigial organs. In all, his bibliography contains 175 titles.

Wyman was a member of all prominent scientific societies and a corporate member of the NAS (1863), although for some unknown reason he later resigned that position.

Daniels, George H. 1968. *American Science in the Age of Jackson* (New York), pp. 227–228. (Contains recent sources.)

Packard, Alpheus S. 1886. *NAS Biog. Mem.* **2:** pp. 75–1 6 (Contains Wyman's complete bibliography.)

Young, Ira (1801–1858). Dartmouth College, 1833–1858.

Ira Young, born in Lebanon, New Hampshire, graduated

from Dartmouth College in 1828. He spent two years teaching in academies, during which time he may have studied theology, and then returned to Dartmouth as tutor. In 1833 he was made professor of mathematics and natural philosophy and in 1838 became professor of astronomy and natural philosophy. Young, an efficient if somewhat uninspiring teacher, was an exceptionally capable organizer and administrator. He succeeded in convincing Dr. Shattuck to donate the money for the Dartmouth College Observatory; and his December 23, 1852, letter to Shattuck (in Dartmouth Archives) certainly indicates that he was familiar with the uses of scientific instruments, which he spent five months of 1853 purchasing in Europe. The nineteenth-century Dartmouth chronicles are full of praise for Young's erudition, which is difficult to evaluate. It is clear, however, that his entire adult life was devoted to the study of science, even if he himself was not a contributor.

Chapman, George T. 1867. *Sketches of the Alumni of Dartmouth College* (Cambridge, Mass.), p. 246.

Lord, John King. 1913. *A History of Dartmouth College* (Concord, New Hampshire), pp. 294–300.

Lord, Nathan. 1859. *A Discourse Commemorative of Ira Young* (Hanover, New Hampshire).

Smith, Baxter Perry. 1878. *History of Dartmouth College (Boston), pp. 290–297.*

Index